Series/Number 07–145

T0222726

AN INTRODUCTION TO GENERALIZED LINEAR MODELS

GEORGE H. DUNTEMAN

MOON-HO R. HO
Department of Psychology, McGill University, Montreal, Quebec, Canada

Division of Psychology, Nanyang Technological University, Singapore

SAGE PUBLICATIONS
International Educational and Professional Publisher
Thousand Oaks   London   New Delhi

For information:

Sage Publications, Inc.
2455 Teller Road
Thousand Oaks, California 91320
E-mail: order@sagepub.com

Sage Publications Ltd.
1 Oliver's Yard
55 City Road
London EC1Y 1SP
United Kingdom

Sage Publications India Pvt. Ltd.
B-42, Panchsheel Enclave
Post Box 4109
New Delhi 110 017 India

Printed in the United States of America

Library of Congress Cataloging-in-Publication Data

Dunteman, George H. (George Henry), 1935–2004
An introduction to generalized linear models / George H. Dunteman, Moon-Ho R. Ho.
    p. cm.—(Quantitative applications in the social sciences; 145)
Includes bibliographical references and index.
ISBN 0–7619–2084–6 (pbk.)
    1. Regression analysis—Mathematical models. 2. Linear models (Statistics) I. Ho, Moon-Ho R. II. Title. III. Sage university papers series. Quantitative applications in the social sciences; 145.
HA31.3.D86 2006
519.5′36—dc22   2005012705

This book is printed on acid-free paper.

14   15   16   17   18   9   8   7   6   5   4   3

Acquisitions Editor:	Lisa Cuevas Shaw
Editorial Assistant:	Karen Gia Wong
Production Editor:	Melanie Birdsall
Copy Editor:	Daniel Hays
Typesetter:	C&M Digitals (P) Ltd
Proofreader:	Chloe Kristy
Indexer:	Naomi Linzer

Rosarie, George E., Elizabeth, and Alyssa
—G. D.

To my parents for support, patience, and endurance
—M. H.

CONTENTS

LIST OF FIGURES AND TABLES

SERIES EDITOR'S INTRODUCTION

The course of editing this book has taken an unusual path: A change in authorship as well as editorship took place. My predecessor, Michael Lewis-Beck, was wise in seeing the value of adding to the series an introductory title on the generalized linear model. He saw through the editing of the prospectus and earlier drafts of the manuscript before stepping down as editor in early 2004. Sadly, George H. Dunteman passed away right after completing what he thought was a final draft. Further revisions were completed by Moon-Ho R. Ho, who gallantly took up the challenge and brought the manuscript to fruition with important additions and revisions to the original draft.

The outcome variables that social scientists analyze can be continuous or discrete. In our series, we have many titles that deal with the type of models represented by the classical linear regression that requires a continuous dependent variable (and a number of crucial assumptions). When the dependent variable is noncontinuous, often the probability of event occurrence is the object of a statistical model, but it can also be frequency or log frequency. During the past two decades, various forms of logit and probit (and log linear) models have become a standard issue in the social scientist's methods repertoire and the topic of quite a few titles in the series.

The relation between the two types of models—those for continuous outcome variables and those for discrete dependent variables—becomes transparent in the framework of the generalized linear model. In the social sciences, researchers are familiar and comfortable with linear or linearizable independent variables on the right-hand side of the equation, expressed as a linear combination of x and β. The dependent variable y on the left-hand side in the two types of models, however, may take on various forms, including metric, binary, ordinary, multinomial, and count. The random outcome of y in the two types of models may be distributed according to the normal, the binomial, the Poisson, the gamma distributions, among others, and all these distributions belong to the exponential family of distributions. Once we have made the proper assumption of the random distribution in y following the exponential form, the remaining task is to specify the link between the expectation of the random variable y and linear combination of x and β. This mapping of the expected random outcome variable y to the linear combination of x and β is part and parcel of the generalized linear model.

So far, we have two titles specifically discussing the generalized linear model: Gill's *Generalized Linear Models: A Unified Approach* (No. 134) and Liao's *Interpreting Probability Models: Logit, Probit, and Other Generalized Linear Models* (No. 101). The former presents the generalized linear model systematically and slightly more theoretically, and the latter provides a unified method for interpretation of estimation results from generalized linear models. The current book, however, has a more humble but nevertheless more down-to-earth goal: For the rank-and-file social science researchers who have mastered classical linear regression, how do they move from the linear regression model to the other type of models for noncontinuous dependent variables without losing sight of the common roots and similarities of the two types of models? The authors walk the reader through such process and enlighten the uninitiated about generalized linear models along the way, thus providing a good addition to the series.

—*Tim Futing Liao*
Series Editor

ACKNOWLEDGMENTS

Sage and the authors thank the reviewers of this text for their invaluable contribution.

AN INTRODUCTION TO GENERALIZED LINEAR MODELS

George H. Dunteman

Moon-Ho R. Ho
Department of Psychology, McGill University,
Montreal, Quebec, Canada

Division of Psychology, Nanyang Technological University,
Singapore

1. GENERALIZED LINEAR MODELS

Generalized linear models, as the name implies, are generalizations of the classical linear regression model. The classical linear regression model assumes that the dependent variable is a linear function of a set of independent variables, and that the dependent variable is continuous and normally distributed with constant variance. The independent variables can be continuous, categorical, or a combination of both. Multiple regression, analysis of variance, and analysis of covariance are examples of classical linear models. They can all be written in the form $y = \beta_0 + \sum_{j=1}^{p} \beta_j X_j + \varepsilon$, where y is the continuous dependent variable, X_j's are the independent variables, and ε is assumed to be a normally distributed error. The dependent variable y is decomposed into two components, a systematic component $\beta_0 + \sum_{j=1}^{p} \beta_j X_j$ and an error component ε. The systematic component is the expected value of y, $E(y)$, for a given set of values for the X_j's. The expected value of y, $E(y)$, is the mean of y, μ_y, for a given set of values for the independent variables, the X_j's; that is, $E(y|X_1, \ldots, X_p) = \beta_0 + \sum_{j=1}^{p} \beta_j X_j$. It is a conditional mean that depends on the values of the X_j's. The goal of regression analysis is to find a set of independent variables that have high explanatory power as measured through goodness of fit. This means that we can explain a large part of the variation in y by a linear combination of the independent variables. If the regression parameters, the β_j's, are large, then as the values of the X_j's change from observation to observation, the expected value of y or the conditional mean of y will vary considerably. If this variation in the conditional mean or predicted value is large relative to the variation in ε, then we have a

1

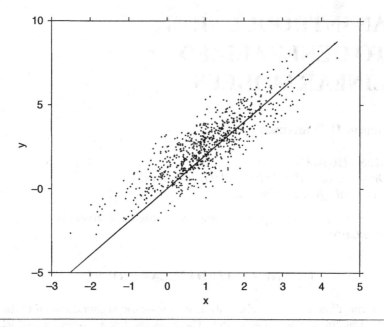

Figure 1.1   Linear Regression Model

useful model for predicting future values of y for given values of the independent variables and for understanding the relative importance of the different independent variables in explaining the variation in the dependent variable y. Figure 1.1 shows a simple linear regression model (with $\beta_0 = 1$ and $\beta_1 = 1.5$). We estimate the regression parameters, β_j's, by collecting measurements of y, X_1, X_2, ..., X_p on a random sample of observational units. For our purposes, the observational units are usually people, but in other applications the units could be anything, such as trees, cows, or even rivers. If we index the people by i and the variables by j, then we can estimate the β_j's by minimizing the error sum of squares

$$\sum_{i=1}^{n} \left(y_i - \beta_0 - \sum_{j=1}^{p} \beta_j X_{ij} \right)^2.$$

Here, subscript i is added to emphasize the fact that the values of the independent variables vary from subject to subject. This method of regression parameter estimation is commonly known as ordinary least squares.

This linear regression model has served the social sciences as well as the other sciences extremely well since its initial development in the 19th century. It is easily formulated, easy to understand, and the regression coefficients are easily estimated by ordinary least squares. Because of these factors, it is still in wide use today across all the sciences. Although it assumes normally distributed errors, it is robust when the errors are only approximately normally distributed.

Nevertheless, it has become increasingly recognized during the past several years that the linear regression model has limitations. It assumes that the dependent variable is continuous or at least quasi-continuous, such as achievement test scores, measures of personality traits, and so on. It also assumes that the continuous variable is at least approximately normally distributed and that its variance is not a function of its mean. Generalized linear models were introduced by Nelder and Wedderburn (1972) to address those limitations. Generalized linear models are a family of models developed for regression models with nonnormal dependent variables.

In many applications, the dependent variable is categorical or consists of counts or is continuous but nonnormal. An example of a categorical dependent variable is a binary variable that takes on only two discrete values, 0 or 1, where 1 indicates the occurrence of an event (e.g., dropping out of college) and 0 the nonoccurrence of an event (e.g., not dropping out of college). The goal is to model the probability of the occurrence of the event of interest. It will be shown later that logistic regression, a type of generalized linear model, is appropriate for this type of data.

An example of a dependent variable involving counts is the number of drug abuse treatment episodes in a 5-year period for a population of substance abusers. Again, it will be shown that Poisson regression, another type of generalized linear model, is appropriate for this situation. In both these cases, the dependent variable is not continuous and is far from being normally distributed. Also, 0-1 binary and count variables are nonnegative, whereas continuous dependent variables in regular regression can take on both positive and negative values.

An example of a nonnormal continuous distribution that has many applications is the gamma distribution. The gamma distribution is skewed, takes on only positive values, and its variance is a function of its mean. It is used to model a wide variety of dependent variables that can take on only positive values, such as income, survival time, and amount of rainfall. Models with gamma distributed dependent variables can be modeled within a generalized linear model framework.

It should be noted that the independent variables can take on a wide variety of distributional forms for a given distribution on the dependent variable, and they are not limited to the same distribution as the dependent variable. For example, the independent variables associated with a normally distributed dependent variable can exhibit a wide variety of nonnormal distributions, such as uniform or multimodal. As mentioned previously, regular regression assumes that whereas the mean of y varies with the independent variables, the variation of ε about the conditional means remains constant. For binary variables and count variables, the variation about the conditional mean is a function of the mean. For binary variables, the conditional mean of the dependent variables is a probability (p) (e.g., the probability of the occurrence of 1, the event), and the variation of the 0's and 1's about this mean is $p(1 - p)$, which is a function of the mean (p). Because p, the mean, varies as a function of the independent variables, so does the variance of the binary variable. For count variables, the Poisson distribution is frequently used, and for this distribution the variance is equal to its mean. Therefore, as the conditional mean of the Poisson distribution varies as a function of the independent variables, so does its variance. Generalized linear models, in this case the logistic and Poisson regression models, explicitly incorporate the relationship of the mean and variance through their probability distributions in the formulation of the model and the estimation of its regression parameters.

Classical regression also assumes that the model is linear in the regression parameters. That is, it is assumed that the expected value or conditional mean is a linear function of the regression parameters. For example, $E(y|X_1, X_2) = \beta_0 + \beta_1 X_1 + \beta_2 X_2$ or even $\beta_0 + \beta_1 X_1 + \beta_2 X_2 + \beta_3 X_1^2 + \beta_4 X_2^2 + \beta_5 X_1 X_2$. Note that the second model is linear in the parameters but nonlinear in the independent variables. In fact, classical linear regression is a specific case of a generalized linear model in which the conditional mean of the dependent variable is modeled directly rather than some transformation of the conditional mean. For other generalized linear models, the conditional mean cannot be written as a linear function of the regression parameters, but some nonlinear function of the conditional mean can be written as a linear function of the parameters; hence the name generalized linear models.

A simple example of a generalized linear model is the Poisson regression model (Figure 1.2). All the characteristics of a generalized linear model can be easily seen in this case. Moreover, it is easy to see the contrasts between this generalized linear model and a classical linear model.

In the case of Poisson regression, the expected value or conditional mean of the Poisson distributed dependent variable is

$$\lambda_i = e^{\beta_0 + \sum_{j=1}^{p} \beta_j X_{ij}}.$$

Here, λ_i is the conditional mean of the Poisson distribution for an individual i. It is conditional in that the mean depends on the regression parameters, the β_j's, which are constant, and the specific values of the X_j's, which vary over the units of analysis (e.g., the individuals). We can compute the conditional mean λ_i for individual i by substituting his or her values of the independent variables, the X_{ij}'s, where X_{ij} is the value of the jth independent variable for individual i. In order to do this, we must have estimated the regression parameters

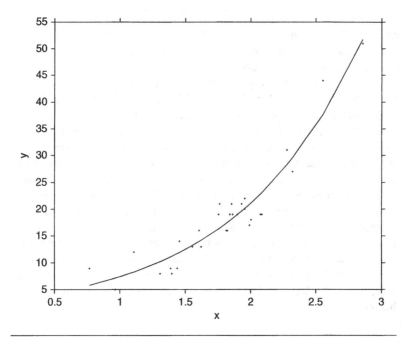

Figure 1.2 Poisson Regression Model

(the β_j's), which are unknown constants. How this is done is discussed later. We need to use the maximum likelihood method instead of least squares.

When the distribution of the dependent variable is nonnormal and its variance is a function of its mean, least squares estimates are no longer equal to maximum likelihood estimates as they are for the normal distribution. In these cases, the likelihood function must be expressed in terms of the appropriate probability density to obtain both proper parameter estimates and their standard errors. Using least squares would result in both erroneous parameter estimates and their standard errors.

The main point is that the conditional mean is not a linear function of the β_j's. If we take the natural logarithm of both sides of the Poisson regression model above, we obtain $\log_e(\lambda_i) = \beta_0 + \sum_{j=1}^{p}\beta_j X_{ij}$. We have linearized the relationship between the Poisson distributed dependent variable and the independent variables by performing a nonlinear transformation on the conditional mean, λ (i.e., $\log_e(\lambda)$). We shall see that $\log_e(\lambda)$ is called the canonical link function for the Poisson regression model. It transforms the conditional mean λ of the dependent variable such that the transformed value, $\log_e(\lambda)$, is a linear function of the regression parameters. It is called canonical because $\log_e(\lambda)$ is the natural parameter of the Poisson distribution when it is expressed in exponential form. We shall also see later that the variance of a Poisson variable is equal to its mean so that if the conditional mean of the Poisson distribution increases, then so does the conditional variance associated with the conditional mean.

There are several good books on generalized linear models (Fahrmeir & Tutz, 1994; Le, 1998; McCullagh & Nelder, 1989; McCulloch, & Searle, 2001), but they usually assume a relatively high level of statistical sophistication on the part of the reader. This book assumes only basic knowledge of statistical inference and some familiarity with multiple regression. Knowledge of elementary calculus and elementary matrix algebra is not assumed, although they may be helpful in a few sections of the book. Those with little or no background in these subjects may skip or skim over those sections with little or no loss of continuity. This book is written in an informal manner and discusses the relevant statistical concepts in an intuitive manner. Its goals are to inform the reader about different types of data and allow him or her to choose the appropriate statistical model for analyzing the data and interpreting the results. In the appendix, we provide examples of how to use statistical software, SAS (SAS Institute, 2002), to fit the generalized models discussed in this book.

2. SOME BASIC MODELING CONCEPTS

We discuss the fundamental concepts of statistical modeling in the context of regular multiple regression analysis. It assumes a continuous distribution for the dependent variable with constant variance for each observation. It also assumes that the predicted value of y, its conditional mean, is a linear function of the regression parameters. We will see later that the regular multiple regression model is one of a number of specific generalized linear models if we assume that the error is normally distributed.

For three independent variables, the model can be written as $y_i = \beta_0 + \beta_1 X_{i1} + \beta_2 X_{i2} + \beta_3 X_{i3} + \varepsilon_i$, where i identifies the observation that in most applications is a person. It is assumed that ε_i has mean 0 and constant variance σ^2. In addition, it is assumed that ε_i is uncorrelated with the independent variables. The systematic component of the model is $\beta_0 + \beta_1 X_{i1} + \beta_2 X_{i2} + \beta_3 X_{i3}$ and is the expected value of y_i or the conditional mean on the dependent variable for the ith observation given the values of X_{i1}, X_{i2}, X_{i3} for the ith observation. We express this as

$$E(y_i | X_{i1}, X_{i2}, X_{i3}) = \mu_i = \beta_0 + \beta_1 X_{i1} + \beta_2 X_{i2} + \beta_3 X_{i3}.$$

The random component of the model is ε_i. We can see that as the independent variables vary, the conditional mean, μ_i, varies. The associated regression parameters, β_1, β_2, and β_3, indicate how strongly each of the independent variables is associated with the dependent variable, y_i. In general, the larger the regression parameter, the stronger the relationship of the associated variable with the dependent variable assuming that the independent variables all have approximately the same variance. The β's are constant from observation to observation. It is the X's that vary from observation to observation and affect the changes in the conditional means, the μ_i's. The parameter β_0 is called the intercept. It is the y-intercept of the multidimensional space with four variables, the expected value of the dependent variable when all the independent variables are set to 0.

The independent variables can be continuous, categorical, or a combination of both. They can also be transformations of the X's, such as X^2 and $\log X$, and interactions such as $X_1 X_2$, as long as the model can be written in a form that is linear in the parameters; for example,

$$E(y|X_1, X_2) = \mu_{y|X_1, X_2, X_1^2, X_2^2, X_1 X_2}$$
$$= \beta_0 + \beta_1 X_1 + \beta_2 X_2 + \beta_3 X_1^2 + \beta_4 X_2^2 + \beta_5 X_1 X_2.$$

Categorical Independent Variables

Categorical independent variables can be represented by the use of indicator variables as described later. The indicator variables define contrasts between the levels of the categorical independent variables with respect to the level of the dependent variable. These contrasts are specified to avoid redundant information in the $X'X$ matrix that would make it singular and, hence, not invertible. We begin with the case of categorical variables with two levels and then generalize to categorical variables with an arbitrary number of levels. An example of a two-level categorical independent variable is a drug abuse treatment variable defined with two levels— treatment and no treatment. We can quantify this variable by the use of a 0-1 indicator variable, where 1 indicates that the individual has undergone treatment, and 0 indicates that the individual has not undergone treatment. If this is the only independent variable and y is a normally distributed measure of, for example, belief in the efficacy of treatment for drug abuse for the population of drug abusers, then the conditional mean of y can be expressed as $E(y|X_1) = \beta_0 + \beta_1 X_1$, where $X_1 = 1$ if the individual has undergone treatment or 0 otherwise. For those who have undergone treatment, $E(y|X_1=1) = \beta_0 + \beta_1(1) = \beta_0 + \beta_1$, and for those who have not undergone treatment, $E(y|X_1=0) = \beta_0 + \beta_1(0) = \beta_0$. Thus, β_1 is a contrast in the conditional mean for those who have had treatment with those who have not had treatment. β_1 is sometimes called the effect of X_1 on y, which in this case is treatment. The group coded 0 is called the reference group. It is arbitrary which group is coded 1. Note that only one indicator variable is required to denote which of two categories the individual belongs.

This use of indicator variables can be extended to categorical variables that have three or more levels or categories (e.g., racial groups classified as Caucasian, African American, Asian, and Other—a categorical variable with four categories). Here, there are four possible indicator variables: $X_1 = 1$ if Caucasian, 0 otherwise; $X_2 = 1$ if African American, 0 otherwise; $X_3 = 1$ if Asian, 0 otherwise; and $X_4 = 1$ if other racial category, 0 otherwise. Only three of the four indicator variables are needed to determine an individual's racial category. (If we know the values of three of the four indicator variables for an individual, then the value of the remaining indicator variable is automatically given.) If we included all four indicators in a regression model, then, as a set, they would be redundant, and their associated regression parameters could not be estimated from a sample of individuals. Which racial category indicator variable to be thrown out is arbitrary and becomes the reference category for interpreting the regression parameters associated with the remaining three indicator variables.

Let us say that we want to examine the effect of race on a normally distributed achievement test score, where race is defined by the categorical variable discussed previously. Let us arbitrarily discard X_4, the indicator variable for the Other racial group. The Other racial group becomes the reference group. Then, we can express the conditional mean of y, conditional on the racial category, as

$$E(y|X_1, X_2, X_3) = \mu_{y|X_1,X_2,X_3} = \beta_0 + \beta_1 X_1 + \beta_2 X_2 + \beta_3 X_3.$$

For Caucasians, $X_1 = 1$, $X_2 = 0$, and $X_3 = 0$ so that the expected value for Caucasians is $\mu_{y|X_1=1, X_2=0, X_3=0} = \beta_0 + \beta_1(1) + \beta_2(0) + \beta_3(0) = \beta_0 + \beta_1$. Similarly, the conditional mean of y for African Americans is $\mu_{y|X_1=0, X_2=1, X_3=0} = \beta_0 + \beta_2$, and the conditional mean of y for Asians is $\mu_{y|X_1=0, X_2=0, X_3=1} = \beta_0 + \beta_3$. Because the Other racial category has a 0 for each of the three indicators, the conditional mean is $\mu_{y|X_1=0, X_2=0, X_3=0} = \beta_0$. Thus, the difference between the Caucasian conditional mean, $\beta_0 + \beta_1$, and the Other racial group conditional mean, β_0, is β_1. Thus, β_1 is a contrast between the Caucasian mean and the Other race mean, the reference category. Similarly, β_2 and β_3 are the differences between the African American and Asian means and the Other race mean, respectively. Note that we can obtain any other contrasts between the racial groups by taking various differences between β_1, β_2, and β_3. For example, the difference between the African American mean and the Caucasian mean is $(\beta_1 - \beta_2)$. Note that in this model, there are four racial subpopulations and four regression parameters: β_0, β_1, β_2, and β_3. For models with categorical independent variables, there can never be more regression parameters than there are cells defined by crossing all the categorical variables. In our example, we had one categorical race variable with four levels or cells.

Let us take another slightly more complicated regression model with two categorical independent variables, each with two levels. Let us suppose that the dependent variable, y, is a normally distributed variable measuring satisfaction with treatment for drug abusers. The independent variables are the indicator variables X_1, which is 1 if the patient is in a residential program and 0 if the patient is in an outpatient program, and X_2, which is 1 if the patient is male and 0 if the patient is female. A main effects model could be expressed as $\mu_{y|X_1,X_2} = \beta_0 + \beta_1 X_1 + \beta_2 X_2$, where β_1 is the contrast in the conditional mean of the residential clients versus the outpatient clients. Similarly, β_2 is the effect of gender, the contrast between males and females. Both these contrasts, the parameters β_1 and β_2, are adjusted for the effects of the remaining variable. That is, β_1 is adjusted for the effect of X_2, and β_2 is adjusted for the effect of X_1.

We could enhance this model by including a parameter reflecting the interaction between gender and the two category treatment variables. This type of interaction variable is formed by taking the cross-product of the two indicator variables. This results in a third variable, $X_1 X_2$, which is also a 0-1 variable that takes the value 1 for clients who are both male and in a residential treatment program and 0 otherwise. The model then becomes $\mu_{y|X_1,X_2,X_1X_2} = \beta_0 + \beta_1 X_1 + \beta_2 X_2 + \beta_3 X_1 X_2$. Now the effect of X_1 on y depends on the level of X_2. If $X_2 = 0$, then $\beta_3 X_1 X_2 = 0$, and the effect of X_1 is β_1 because the only term involving X_1 is $\beta_1 X_1$. If $X_2 = 1$, however, then the effect of X_1 on y is $(\beta_1 + \beta_3)$ because the term $\beta_1 X_1 + \beta_3 X_1 = (\beta_1 + \beta_3) X_1$ is in the model.

Essential Components of Regression Modeling

Here, we summarize the essential components of regression modeling. First, we select a dependent variable y that we hypothesize is a function of a set of independent variables. The independent variables selected depend on the objectives of the research and the researcher's substantive knowledge of the domain of investigation. A statistician cannot provide much help here. A statistician, however, can advise on the study design, especially the size of the sample, the probability model used to select the study participants, and the type of generalized linear regression model.

Then, we want to compare the fit of various models by comparing the corresponding deviances until we find a model that contains a relatively small set of conceptually appealing independent variables and provides a good fit to the dependent variable. The deviance is a measure of the goodness of fit of the regression model. Each generalized linear model has a specific deviance associated with it. For the regression model based on the normal distribution, it is the familiar error sum of squares. Deviance is discussed in more detail subsequently.

Usually, our initial set of independent variables contains a subset of independent variables that have statistically insignificant regression parameters associated with them. By comparing the fit of various models, we reduce the number of parameters in the model. Once this model is identified, then our interest focuses on the regression parameter estimates and their estimated standard errors. The regression parameter estimates indicate the relative importance of the independent variables in explaining the variation in the dependent variable.

In summary, for generalized linear models, we need to specify the probability distribution of the dependent variable. So far, we have discussed the normal and Poisson distribution, but there are other distributions such as the binomial, which we discuss in detail later. The other critical component that we need to specify is the form of the regression equation, which indicates how the conditional mean is related to the independent variables. We discussed the form of the regression equations for the dependent variable under normal and Poisson regression, and we will discuss other forms in subsequent chapters. The specification of the error distribution in terms of a probability distribution, such as the normal, binomial, Poisson, and so on, and the form of the regression equation is needed to specify the appropriate log-likelihood function. The log-likelihood function is identical to the log of the probability density function except that in the former case the sample data are considered fixed and the parameters are the variables, whereas in the latter case the parameters are considered fixed and the data vary. The log likelihood for the sample is then used to obtain the maximum likelihood regression parameter estimates and their estimated standard errors. The maximum likelihood estimates, except for the normal regression model, cannot be solved by standard analytic methods because of the complexity of the system of equations. Computer algorithms such as iterative reweighted least squares are required for parameter estimation.

3. CLASSICAL MULTIPLE REGRESSION MODEL

Multiple regression analysis is a generalized linear model in which the conditional mean is a linear function of the regression parameters. This is in contrast to other generalized linear models in which some function of the conditional mean is a linear function of the regression parameters. That function of the mean, $g(\mu)$, is called a link function. For example, the link function for the Poisson distribution is $\log_e(\mu)$, which in turn is expressed as a linear function of the regression parameters—that is, $\log_e(\mu) = \beta_0 + \sum_{j=1}^{p} \beta_j X_j$. For the normal distribution, the link function is the identity function, $g(\mu) = \mu$. The multiple regression model can be written as

$$E(y|X_1, X_2, \ldots, X_p) = \mu = \beta_0 + \sum_{j=1}^{p} \beta_j X_j.$$

If we include a column vector of one's associated with the intercept parameter β_0, then the model can be expressed more compactly as

$$E(y|X_0, X_1, X_2, \ldots, X_p) = \mu = \sum_{j=0}^{p} \beta_j X_j,$$

where $X_0 = 1$. The error, $y - \mu = \varepsilon$, is assumed to be normally distributed with constant variance.

We begin the discussion of generalized linear models by presenting an example of multiple regression analysis taken from Chatterjee and Price (1977). It is based on survey data collected on clerical employees of a large financial organization. The survey contained questions concerning employees' satisfaction with their supervisors. One question was a measure of the overall performance of their supervisor. Other questions covered specific activities involving the interaction between the employee and his or her supervisor. The goal of the study was to explain the relationship between specific supervisor characteristics and overall satisfaction with supervisors as perceived by the employees.

In this example, six questionnaire items were chosen as possible explanatory variables. The dependent variable was the overall rating of the job being done by the employee's supervisor. The rating was based on a 5-point scale ranging from 1 (very satisfactory) to 5 (very unsatisfactory). The six independent variables were also based on rating the supervisor's behaviors on a 5-point scale. They were X_1, handles employee complaints; X_2, does not allow special privileges; X_3, opportunity to learn new things; X_4, raises based on performance; X_5, too critical of poor performance; and X_6, rate of advancing to better jobs. These independent variables can be grouped as follows: One group is composed of variables X_1, X_2, and X_5, which concern direct interpersonal relationships between the employee and the supervisor. Another group is composed of two variables, X_3 and X_4, which concern the job as a whole in contrast to interpersonal relationships. The remaining variable, X_6, does not involve the evaluation of the supervisor but is a general measure of the employee's perception of his or her progress in the company. The data were collected from 30 departments that were randomly selected from all the departments in the organization. Each department had approximately 35 employees and one supervisor.

Assumptions and Modeling Approach

Note that the dependent variable could only take on five values, and the distribution across the five values was skewed in that the supervisors were more likely to be rated favorably rather than unfavorably. This violates the common assumption of multiple regression analysis that the dependent variable is a continuous variable that is normally distributed. The model is robust against the departure from normality, however, as long as there is a relatively large number of values for the dependent variable and the distribution is relatively symmetrical.

The basic multiple regression model also assumes that the responses on the dependent variable across the observations—in this case, the employees— are independent of one another. This assumption may also be violated in that employees within a department may respond somewhat similarly to one another. That is, employee responses on the dependent variable within a department may be correlated because they are subject to a set of common influences associated with their department. We would not expect the dependent variable responses from employees in different departments to be correlated, however. The correlation of responses across individuals within a group or cluster is called an intraclass correlation. The basic multiple regression model can be modified to account for the intraclass correlation in the estimation of regression parameters and their standard errors. The regression model is modified by adding random components to the model and is referred to as the mixed effects model. This class of models will not be discussed further in this book.

It is important to note that no assumptions are made concerning the distributions of the independent variables. They can be continuous, discrete, highly skewed, and correlated with one another. Thus, there is no problem with the independent variables in our supervisory performance example, even though they are discrete and, possibly, highly skewed. This is true for any regression model.

In our example, there are problems with the distribution of the dependent variables being nonnormal and the presence of the intraclass correlation among the responses of employees within the same department. Chatterjee and Price (1977), however, ameliorated these two problems in their example by aggregating the individual-level data within each of the 30 departments and using the department as the unit of analysis in their multiple regression model. This makes sense because they were interested in predicting the overall performance of the department supervisors. There was one supervisor for each of the 30 departments, and it makes sense to aggregate the approximately 35 employees' perceptions for each of the six independent variables and the dependent variable within each supervisor's department to predict the supervisor's performance. This procedure is much more reliable than simply having a few employees in each department rate their supervisor's overall performance. This is because employee biases and rating errors tend to cancel out for a large sample of raters. It should be noted, however, that aggregating the data within departments results in information loss concerning the variability of individual responses within departments.

The authors used the following procedure to aggregate the individual-level data to the department level. A dichotomous variable was created for each of the seven items, reflecting the six independent variables and the dependent variable, by collapsing the 5-point response scale into two categories, favorable and unfavorable responses. The favorable responses were ratings 1 (very satisfactory) or 2 (satisfactory), and the unfavorable responses were the remaining responses of 3, 4, or 5. The proportion of favorable responses on each of the seven items was calculated for each of the 30 departments. Thus, we have a column vector of 30 department proportions of favorable overall ratings of supervisor job performance for the dependent variable and a 30 (rows) by 6 (columns) matrix of the proportions of favorable responses on the six independent variables for the 30 departments. These are the data on the 30 departments, the unit of analysis, that are used to estimate the six regression parameters and their associated standard errors.

We will see later that because the dependent variable is the probability of a favorable overall rating of supervisory performance, logistic regression analysis could also have been used to model the relationship of the six independent variables with supervisory performance. We will see that logistic regression is used to model the probability of a particular response occurring—in this case, a favorable supervisory rating. The logistic function is nonlinear. Most of the departmental favorable supervisory rating proportions (probabilities), however, were in the .40 to .60 range, where the logistic function, in many instances, is approximately linear. Thus, the multiple regression model, which is easier to interpret than the logistic regression model, is most likely a good choice for modeling these data.

Results of Regression Analysis

The results of the multiple regression analysis are presented in Table 3.1, which indicates that only X_1 (satisfaction with how supervisor handles employee complaints) is highly significant. It also has the largest regression

TABLE 3.1

Regression Parameter Estimates for Predicting the Proportion
of Favorable Supervisory Ratings

Variable	Regression Parameter	Standard Error	t ratio	Significance Level
X_1 (complaints)	.613	.1610	3.81	.001
X_2 (privileges)	−.073	.1357	−.54	NS
X_3 (learning)	.320	.1685	1.90	.07
X_4 (raises)	.081	.2215	.37	NS
X_5 (criticism)	.038	.1470	.26	NS
X_6 (job advancement)	−.217	.1782	−1.22	NS
Intercept	10.787	11.5890	0.93	NS

NOTE: $R^2 = .7326$; standard deviation of residuals $= 7.068$; NS $=$ not significant at the .05 level.

parameter estimate. The regression parameter estimate, .613, indicates that a one-unit increase in the percentage of employees who are satisfied with the way in which their supervisor handles complaints results in an increase of .613 of a percentage point in the percentage of favorable overall supervisory ratings. That is, an increase of 1% in X_1 results in an increase of .613% in y. An increase of 10 percentage points in X_1 would result in an increase of 6.13 percentage points in y. This indicates a strong relationship between X_1 and the dependent variable when holding the remaining five independent variables constant. The second highest regression parameter was .320 for X_3 (satisfaction with opportunity to learn new things). Although it was statistically significant at the .10 level, it was not significant at the traditional .05 level.

Multiple Correlation

The squared multiple correlation (R^2), which indicates the proportion of explained variation in the dependent variable by the independent variables, was .7326. It is defined as

$$R^2 = 1 - \frac{\text{error sum of squares}}{\text{total sum of squares}},$$

where the error sum of squares (*ESS*) is defined as $\sum_{i=1}^{n}(y_i - \hat{y}_i)^2$, where $\hat{y}_i$ is the predicted value of y_i based on the regression model; that is, $\hat{y}_i = \hat{\beta}_0 + \sum_{j=1}^{p}\hat{\beta}_j X_{ij}$. The total sum of squares (*TSS*) is defined as $\sum_{i=1}^{n}(y_i - \bar{y})^2$, where $\bar{y}$ is the mean of the dependent variable. If *ESS* is small relative to *TSS*, then R^2 will be high. The multiple correlation, $R = \sqrt{.7326} = .856$, can also be defined as the correlation between y and the composite $\hat{y} = \hat{\beta}_0 + \sum_{j=1}^{p}\hat{\beta}_j X_j$.

Testing Hypotheses

A standard test that is usually applied initially is to test the null hypothesis that all the regression coefficients are zero. For our example, the null hypothesis is that $\beta_1 = \beta_2 = \beta_3 = \beta_4 = \beta_5 = \beta_6 = 0$. Note that we do not include the intercept parameter, β_0, in our test because we are interested in only the relationship of the six independent variables with the dependent variable.

To test the null hypothesis that a set of regression parameters is equal to zero, we begin by estimating the error sum of squares for two regression models. One model, called the full model, includes the full set of regression parameters—in our case, all six independent variables. The other model, called the reduced model, omits the independent variables that are hypothesized to have parameters equal to zero. The reduced model for our overall test that there is no relationship between our six independent variables and supervisory ratings is to remove all six independent variables and include only the intercept in the reduced model. The intercept for the reduced model is the mean of the dependent variable, $\bar{y}$. Next, we compute the error sum of squares for the full model, $ESS(FM)$, and the error sum of squares for the reduced model, $ESS(RM)$; compute the difference $ESS(RM) - ESS(FM)$; and divide the difference by the difference in the number of regression parameters for the full and reduced models, denoted by p_f and p_r, respectively. In our case, $p_f = 7$, and $p_r = 1$. This ratio,

$$\frac{ESS(RM) - ESS(FM)}{p_f - p_r},$$

is the numerator of an F ratio. $ESS(RM)$ will be at least as large as $ESS(FM)$ because fewer independent variables are used to predict y—in our case, none. If this positive difference is small, it suggests that the reduced model fits the data about as well as the full model. We need a term to gauge the difference in this ratio. This term forms the denominator of the F ratio and is simply $ESS(FM)$ divided by its degrees of freedom, which is the sample size, n, minus the number of regression parameters that were estimated in the full model. In our case, $(n - p_f) = 30 - 7 = 23$.

Thus, the F-ratio is

$$\frac{\dfrac{ESS(RM) - ESS(FM)}{p_f - p_r}}{\dfrac{ESS(FM)}{n - p_f}},$$

with $(p_f - p_r)$ degrees of freedom in the numerator and $(n - p_f)$ degrees of freedom in the denominator. Under the null hypothesis that the parameters left out of the reduced model are zero, F is distributed with the previously discussed degrees of freedom for the numerator and denominator, respectively. We refer the value of F computed from our data to an F table with the appropriate numerator and denominator degrees of freedom. Chatterjee and Price (1977) computed the F associated with the null hypothesis that all the regression parameters are equal to zero or, equivalently, there is no linear relationship between the six independent variables and overall supervisory performance ratings.

For this F test, the components of the F test were $ESS(FM) = \sum_{i=1}^{30} (y_i - \hat{y}_i)^2 = 1149$; $ESS(RM) = \sum_{i=1}^{30} (y_i - \hat{y}_i)^2 = 4297$; $p_f - p_r = 7 - 1 = 6$; and $n - p_f = 30 - 7 = 23$. This gives an F ratio of

$$\frac{\frac{4297 - 1149}{6}}{\frac{1149}{23}},$$

which equals 10.5, with 6 and 23 degrees of freedom for the numerator and denominator, respectively. Referring this value of F to an F table with 6 and 23 degrees of freedom indicates a significance level of .001. Thus, we would reject the null hypothesis that the regression coefficients are all equal to zero and conclude that the full model $E(y|X_1, X_2, \ldots, X_p) = \mu = \beta_0 + \sum_{j=1}^{p} \beta_j X_j$ provides a better fit to the data than the reduced model $E(y) = \mu = \beta_0$.

We could have surmised this because two of the regression parameters were significant, one, X_1, at the .001 level. The authors checked for violations of model assumptions or model misspecification by plotting the standardized residuals (y axis) against the fitted or predicted values (x axis). The standardized residuals are $y_i - \hat{y}_i$ divided by the standard deviation of the errors or residuals—that is,

$$\sqrt{\frac{\sum (y_i - \hat{y}_i)^2}{n - p_f}}.$$

If the model is correctly specified, then the scatter of the standardized residuals should appear to be random with no systematic pattern in the scatter plot. Also, 95% of the residuals should be between -2 and $+2$ or within 2 standard deviations from the mean of the residuals, which is zero by assumption. Their residual plot showed no evidence of model misspecification.

Chatterjee and Price (1977) also plotted the standardized residuals against the most important independent variable, X_1. Again, the scatter plot looked random, with no large standardized residuals and no evidence of any systematic pattern such as curvature in the scatter plot. Curvature could indicate the need for an X^2 term in the regression model.

Returning to our full model, it is quite clear that only variables X_1 and X_3 seem to be important. Therefore, it seems reasonable to test the null hypothesis that $\beta_2 = \beta_4 = \beta_5 = \beta_6 = 0$ to determine if we can simplify the full model.

To test this hypothesis, we use the same F test procedure previously described. We are contrasting the full model $\mu = \beta_0 + \beta_1 X_1 + \beta_2 X_2 + \beta_3 X_3 + \beta_4 X_4 + \beta_5 X_5 + \beta_6 X_6$ against the reduced model $\mu = \beta_0 + \beta_1 X_1 + \beta_3 X_3$.

Therefore, again, we calculate $ESS(FM)$, $ESS(RM)$, $p_f - p_r$, and $n - p_f$ and substitute them into the F ratio formula. We find that

$$F = \frac{(1254.6 - 1149)/4}{1149/23} = .528,$$

which with 4 and 23 degrees of freedom is a very small value of F that is not significant at the .05 level. Thus, we accept the null hypothesis that the four variables X_2, X_4, X_5, and X_6 do not need to be included in the model and accept the simpler reduced model $E(y|X_1, X_3) = \beta_0 + \beta_1 X_1 + \beta_3 X_3$ over the more complex full model $E(y|X_1, X_2, X_3, X_4, X_5, X_6) = \beta_0 + \sum_{j=1}^{6} \beta_j X_j$.

The R^2 for the reduced model was .7080, which is slightly smaller than the R^2 for the full model (i.e., $R^2 = .7326$). The regression parameter for X_1, the most important, in the reduced model is .643 compared to .613 in the full model. Note that all the independent variables in this model were continuous variables. Categorical independent variables characterizing the departments could also have been included in the models. For example, we could have a 0–1 indicator variable that takes the value 1 if the department performs an accounting function and 0 otherwise. We could also include interaction terms in the model by including the appropriate cross-products between the variables that are hypothesized to interact with one another. We must keep in mind, however, that with only 30 observations, we cannot fit very many parameters. Otherwise, we run the risk of overfitting where there are more parameters than the data can support for reasonable inference. In the extreme case, if our regression model included 30 parameters, the model would fit the data perfectly. We have not simplified anything by having as many parameters as observations, however. Besides, there would be no degrees of freedom for the ESS and, consequently, the testing of alternative hypotheses is precluded. Even with a small number of degrees of freedom for $ESS(FM)$, the F test will have low power. Building a good regression model begins with specifying a dependent variable and a set of independent variables that are driven by a well-formulated set of hypotheses that are, in turn, based on knowledge of the subject matter. The researcher then estimates the parameters of the model and assesses its goodness of fit. The initial full model is then refined by eliminating some parameters by hypotheses testing. The regression diagnostics, such as residual plots, may suggest other modifications of the model, such as the inclusion of X^2 terms or interaction terms. If the data are based on time series data, then the residuals could be correlated, and the revised model must allow for correlated residuals in the estimation of the regression parameters and then testing of various hypotheses about them.

4. FUNDAMENTALS OF GENERALIZED LINEAR MODELING

Chapter 3 discussed a generalized linear model with which we are all familiar, the classical multiple regression model. This chapter indicates how this model can be generalized to other situations in which the dependent variable is discrete, nonnormally distributed, and its variance depends on its mean.

Generalized linear models involve predicting the conditional mean or some function of the conditional mean of a dependent variable as a linear function of a set of independent variables or covariates. That is, for each observation or subject, its expected value, or some function of its expected value of the dependent variable, is conditional on the value of its independent variables or covariates. Except for the normal distribution, the error variance for generalized linear models is a function of the mean. For example, a 0–1 binary outcome variable has a mean of π, the proportion of times for which the event signified by 1 occurs, and a variance of $\pi(1 - \pi)$. To estimate the regression parameters and their standard errors, we need to specify a probability distribution for the error term so that we can specify the appropriate likelihood function and use the likelihood function to solve for the regression parameters.

Generalized linear models allow us to work with data for which the mean of the dependent variable is a nonlinear function of the regression parameters and the response variable is not normally distributed. The two components of a generalized linear model are a link function and an error distribution. The link function is a transformation of the mean of the dependent variable such that this transformed variable is a linear function of the regression parameters. For example, the link function for the Poisson regression model is logarithm, $g(\mu) = \log_e(\mu)$, so that the dependent variable, $g(\mu)$, is a linear function of the regression parameter associated with the independent variables. That is, $\log_e(\mu) = \sum_{j=0}^{P} \beta_j X_j$. Note that $g(\mu)$ is a nonlinear function of the regression parameters because exponentiating both sides of the equation leads to

$$\mu = e^{\sum_{j=0}^{P} \beta_j X_j}.$$

The logarithm link function for the Poisson distribution is also called the canonical link because it is the transformation of μ that becomes the canonical parameter θ when the Poisson distribution is expressed in exponential form; that is, $g(\mu) = \theta = \log_e(\mu)$. This is the link function most commonly used in Poisson regression, although other link functions are possible. For example, we could use the noncanonical identity link $g(\mu) = \mu$. In some instances, a noncanonical link may fit a particular data set better than the canonical link. We shall see later that generalized linear models assume that the dependent variable is a member of the exponential family of distributions. Each distribution that is a member of the exponential family has its own canonical parameter θ that is a function of its mean when the distribution is expressed in exponential form. The function $\theta(\mu)$, of course, is different for different members of the exponential family. For the Poisson distribution, we saw that it was $\theta(\mu) = \log_e(\mu)$. For the binomial distribution associated with logistic regression, it is

$$\theta(\mu) = \log_e \frac{\mu}{1 - \mu}.$$

For the normal distribution, $\theta(\mu) = \mu$—that is, the identity link.

The second component of generalized linear models is that, except for the normal distribution, the variance of the dependent variable is a function of its mean. This is a characteristic of distributions that are members of the exponential family, the underlying response distributions of generalized linear models. The variance for the Poisson distribution is $\mathrm{Var}(y) = \mu$, whereas the variance for the binomial distribution is $\mathrm{Var}(y) = \mu(1 - \mu)$. For the normal distribution, the variance is constant. That is, $\mathrm{Var}(y) = \sigma^2$.

Generalized linear models assume that the observations of the dependent variable, $y_1, y_2, \ldots, y_n$, are independent and share the same form of parametric distribution from the exponential family (see the next section for the definition). The means $\mu_1, \mu_2, \ldots, \mu_n$ associated with the observations can differ, but the observations must each be generated from the same probability distribution (e.g., all generated by the Poisson distribution). This implies that the means differ across observations because the generalized regression model assumes that the mean or some nonlinear function of the mean is related to a set of independent variables. That is, we assume that there is a set of $(p + 1)$ regression parameters, $\beta_0, \beta_1, \ldots, \beta_p$, and an associated set of independent variables, $X_1, X_2, \ldots, X_p$, such that the appropriate link function is $g(\mu) = \sum_{j=0}^{p} \beta_j X_j$.

Next, we examine the characteristics of probability distributions from the exponential family on which generalized linear models are based.

Exponential Family of Distributions

Generalized linear models involve probability distributions that can be expressed in exponential form. These distributions are members of the exponential family of distributions. When expressed in exponential form, there is a canonical parameter that is a function of the mean and a variance function that is also a function of the mean. For example, the canonical parameter for the Poisson distribution is $\log_e(\mu)$, and the variance of the distribution is μ.

The normal distribution is typically expressed as

$$f(y|\mu, \sigma^2) = \frac{1}{\sqrt{2\pi\sigma^2}} e^{-\frac{1}{2}\frac{(y-\mu)^2}{\sigma^2}},$$

where μ and σ^2 are the parameters of the distribution, the mean and variance, respectively. Note that the distribution is already partially in exponential form. Using some algebra, it can be expressed as

$$f(y|\mu, \sigma^2) = e^{\frac{(y\mu - \frac{\mu^2}{2})}{\sigma^2} - \frac{1}{2}(\frac{y^2}{\sigma^2} + \log_e(2\pi\sigma^2))}.$$

All distributions of the exponential family can be expressed as

$$f(y|\theta, \phi) = e^{\frac{y\theta - b(\theta)}{\phi} + c(y, \phi)},$$

where θ is called the canonical or natural parameter, which is a function of the mean (μ) of the distribution; $b(\theta)$ is a function of the canonical parameter, which is also a function of the mean because θ is a function of the mean; ϕ is a dispersion parameter that plays a role in defining the variance of y; and $c(y, \phi)$ is a function of the observation and the dispersion parameter. We can determine what these parameters (θ and ϕ) and functions $b(\theta)$ and $c(y, \phi)$ are for the normal distribution by equating the terms of the normal distribution ($f(y|\mu, \sigma^2)$), which is expressed in exponential form in terms of μ and σ^2, and the terms of the exponential form, which is expressed in terms of the canonical parameters θ and ϕ. We find that $\mu = \theta$, $b(\theta) = \theta^2/2$, $\phi = \sigma^2$, and

$$c(y, \phi) = -\frac{1}{2}\left(\frac{y^2}{\sigma^2} + \log_e(2\pi\sigma^2)\right).$$

One important component, θ, is a function of μ, expressed as $\theta(\mu)$, which is called the canonical link function. It links the mean to the canonical parameter, which in turn can be expressed as a linear function of the regression parameters. Another important component of the exponential family of distributions is the variance function, which is the second derivative of $b(\theta)$

signified by $b''(\theta)$. For the normal distribution, the second derivative of $b(\theta) = \theta^2/2 = 1$. The variance of the distribution is $\phi b''(\theta)$, where ϕ is the dispersion parameter, and $b''(\theta)$ is the second derivative of $b(\theta)$. For the normal distribution $\phi = \sigma^2$ and the variance function $b''(\theta) = 1$ so that the variance of a normally distributed variable is simply σ^2. It is a constant and not a function of the mean.

Each member of the family of exponential distributions has its own link function $\theta(\mu)$ and variance function $b''(\theta)$. The variance function can also be expressed in terms of the mean μ because θ is a function of μ. We denote it as $V(\mu)$, which indicates that the variance is a function of μ. Let us examine the Poisson distribution and express it in exponential form so we can determine $\theta(\mu)$ and $b''(\theta)$. The Poisson distribution has one parameter, the mean μ (also commonly expressed as λ). The Poisson distribution written in terms of the mean, as discussed previously, is

$$f(y|\mu) = \frac{\mu^y e^{-\mu}}{y!}.$$

The notation '!' is called factorial, and $y! = y(y-1)(y-2)(y-3)\cdots 1$. In exponential form, the Poisson distribution can be expressed as $f(y|\mu) = e^{y\log_e\mu - \mu - \log y!}$.

Equating this to the canonical exponential form $f(y|\theta, \phi)$, we have $f(y|\theta, \phi) = e^{y\theta - e^\theta - \log_e y!}$ because $\theta = \log_e\mu$, $b(\theta) = e^\theta$, $\phi = 1$, and $c(y, \theta) = \log_e y!$ The canonical link function is $\theta = \log_e(\mu)$, and the variance function is $b''(\theta)$. The variance function expressed as a function of μ is the identity function $V(\mu) = \mu$ because $e^\theta = e^{\log_e\mu} = \mu$. Because $\phi = 1$, the Poisson distribution with mean μ also has variance equal to μ. The Poisson distribution does not involve an unknown dispersion parameter like the normal distribution because it is a constant equal to 1. The principles illustrated here can be applied to other distributions in the exponential family. For all cases, the canonical parameter can be specified as a linear function of the independent variables. That is, $\theta = \sum_{j=0}^{p}\beta_j X_j$.

Classical Normal Regression

For normally distributed dependent variables, we frequently use the identity link—that is, $g(\mu) = \mu$. Note that the canonical parameter for the normal distribution is $\theta(\mu) = \mu$ so that $g(\mu) = \theta(\mu)$. We model μ directly with no transformation of μ. Thus, $y = \beta_0 + \beta_1 X_1 + \cdots + \beta_p X_p + \varepsilon$, the expected value of y or μ is $\beta_0 + \beta_1 X_1 + \cdots + \beta_p X_p$, and ε is assumed to be normally distributed with mean 0 and variance σ^2. The variance of ε does not depend on the mean as it does for other generalized linear models; it is assumed to be constant across all observations. The regression parameters of this model can be interpreted as the additive effect on μ resulting from a one-unit increase in the corresponding independent variable while holding the other independent variables constant.

For other generalized linear models, such as the logistic and Poisson, the regression parameters indicate the additive effect of a one-unit increase in the corresponding independent variables through the link function—that is, some nonlinear transformation of the mean and not the mean.

Logistic Regression

Logistic regression is used to model the probability of an event such as dropping out of college as a function of student characteristics. The canonical link function for logistic regression is the logit, $\log \frac{\pi}{1-\pi}$, where π is the mean of the binary dependent variable or the probability of the event occurring. Then, $\log \frac{\pi}{1-\pi}$ is expressed as a linear function of the independent variables. That is, $\log \frac{\pi}{1-\pi} = \beta_0 + \beta_1 X_1 + \cdots + \beta_p X_p$, where $\beta_0, \beta_1, \ldots, \beta_p$ are the regression coefficients. The error distribution used in the likelihood function is the binomial distribution. For the binomial distribution, the variance is a function of the mean; the variance equals $\pi(1 - \pi)$. Although this is the typical specification for a logistic regression model, other specifications exist. For example, a probit link function can be used, and a beta-binomial error distribution can be used if there is evidence that there is more variation in the binary dependent variable than is accounted for by the binomial distribution.

For the logistic regression model, $\log \frac{\pi}{1-\pi} = \beta_0 + \beta_1 X_1 + \cdots + \beta_p X_p$ so that the effect of β_j on the logit is additive. It is difficult to interpret the magnitude of the additive effect of the odds on the log; thus, we typically exponentiate both sides of the previous equation to get

$$\frac{\pi}{1-\pi} = e^{\beta_0 + \beta_1 X_1 + \cdots + \beta_p X_p} = e^{\beta_0} e^{\beta_1 X_1} \ldots e^{\beta_p X_p}.$$

We can see from this transformation that the exponentiated regression parameters, e^{β_j}, now represent the multiplicative effect of a one-unit increase in X_j on the odds while holding the remaining independent variables constant. The term e^{β_j} represents the odds of the binary outcome occurring at the value of $X_j + 1$ divided by the odds of the outcome occurring at the value of X_j. Note that we are modeling $\frac{\pi}{1-\pi}$ and not π.

Poisson Regression

Poisson regression is used to model the number of specific events occurring over a given period of time, such as the number of automobile accidents during a 5-year period as a function of driver characteristics. For the Poisson regression model, $\log_e \lambda$, the canonical link, rather than λ, is modeled as a linear function of the independent variables. λ is the mean or expected value of the Poisson distribution; it is the expected number of times an event occurs in a given period of time. It is sometimes called a rate or intensity parameter. Therefore, $\log_e \lambda$ is the link function, and it is assumed to be a linear function of the independent variables. For the Poisson distribution, the variance is a function of its mean; the variance is equal to λ, the mean of the distribution. Again, however, we have problems interpreting the additive effect of a one-unit increase in X_j on the log scale. Thus, we transform the additive model $\log_e \lambda = \beta_0 + \beta_1 X_1 + \cdots + \beta_p X_p$ into the multiplicative model $\lambda = e^{\beta_0 + \beta_1 X_1 + \cdots + \beta_p X_p} = e^{\beta_0} e^{\beta_1 X_1} \ldots e^{\beta_p X_p}$ by exponentiating both sides of the log model. Then, e^{β_j} is the multiplicative effect on λ due to a one-unit increase in X_j, the covariate associated with β_j while holding the other independent variables constant.

Proportional Hazards Survival Model

Survival analysis involves modeling the time until an event happens (e.g., death, dropping out of school, or finding a job) or some function of time until an event happens as a function of independent variables. In our case, we will model the hazard, $h(t)$, which is a function of time. The hazard is the instantaneous probability of an event happening at a given time. A widely used model for survival time data is the Cox proportional hazard model, defined as $h(t) = h_0(t) e^{\beta_0 + \beta_1 X_1 + \cdots + \beta_p X_p}$, where $h_0(t)$ is called baseline hazard at time t. It is the hazard function in the absence of covariates. If we divide both sides by $h_0(t)$, we get

$$\frac{h(t)}{h_0(t)} = e^{\beta_0 + \beta_1 X_1 + \cdots + \beta_p X_p},$$

which shows from where the term *proportional* derives. For each individual, $e^{\beta_0 + \beta_1 X_1 + \cdots + \beta_p X_p}$ is constant across time, which shows that at each value of t, any individual's hazard function is a constant proportion of the baseline hazard. Taking logarithm on both sides of the Cox proportional model, the log of the hazard can be modeled as a linear function of the independent variables.

5. MAXIMUM LIKELIHOOD ESTIMATION

Maximum likelihood estimation is based on the conceptually appealing notion that the estimated parameters—in our case, estimated regression parameters—should be those that maximize the value of the density function specified for the sample data. That is, conditional on the sample data, maximum likelihood estimation finds the parameter values that most likely generated the sample observations. The probability density function when specified as a function of the parameters given the data is called the likelihood function. The density function and likelihood function are the same, but the former regards the parameters as fixed and the data varying, whereas the latter regards the data as fixed and the parameters as varying. The maximum likelihood estimates are those parameter estimates that maximize the likelihood function. In some cases, they can be solved for analytically by partial differentiation. In more complex cases, the solutions are not analytically tractable, and computer algorithms must be used. Maximum likelihood estimates have excellent statistical properties, such as efficiency.

Maximum likelihood estimation requires a probability density function to be specified that is assumed to characterize the sample data. Because of the specific mathematical form of the density function for the normal regression model, maximum likelihood estimates of the regression parameters are identical to the least square estimates. This is not true for other generalized linear models.

The regression equations we have discussed have population parameters that need to be estimated from a random sample of individuals. To do this, we need a statistical model indicating how the data were generated. For normally distributed data, the density function of the normally distributed random variable y is

$$f(y|\mu, \sigma^2) = \frac{1}{\sqrt{2\pi\sigma^2}} e^{-\frac{1}{2}\frac{(y-\mu)^2}{\sigma^2}}.$$

The ith sample member's observation on y is y_i and is assumed to be distributed as

$$f(y_i|\mu_i, \sigma^2) = \frac{1}{\sqrt{2\pi\sigma^2}} e^{-\frac{1}{2}\frac{(y_i-\mu_i)^2}{\sigma^2}}.$$

That is, for each observation, y_i, is assumed to be generated from a normal distribution with its own mean, μ_i, but each observation has the constant variance σ^2 because there is no i subscript on σ^2. This meets the normal regression assumption that the conditional means differ from individual to individual as a function of covariates but that the variance of y_i remains constant.

Now, a regression model hypothesizes that μ_i is a linear function of regression parameters and can thus be expressed as $\mu_i = \beta_0 + \sum_{j=1}^{p}\beta_j X_{ij}$. Thus, we can express the density of y_i as

$$f(y_i|\beta_0, \ldots, \beta_p, \sigma^2) = \frac{1}{\sqrt{2\pi\sigma^2}} e^{-\frac{1}{2}\frac{(y_i-(\beta_0+\sum_{j=1}^{p}\beta_j X_{ij}))^2}{\sigma^2}}.$$

To save space, we denote $\beta_0, \ldots, \beta_p$ collectively as a colume vector $\boldsymbol{\beta}$. Since the y_i are assumed to be independent, the joint distribution of the observations making up the sample can be expressed as

$$f(y_1, y_2, \ldots, y_n | \boldsymbol{\beta}, \sigma^2) = f(y_1 | \boldsymbol{\beta}, \sigma^2) f(y_2 | \boldsymbol{\beta}, \sigma^2) \ldots f(y_n | \boldsymbol{\beta}, \sigma^2),$$

where n is the sample size. Thus,

$$
\begin{aligned}
f(y_1, y_2, \ldots, y_n | \boldsymbol{\beta}, \sigma^2) = {} & \frac{1}{\sqrt{2\pi\sigma^2}} e^{\frac{-1}{2} \frac{(y_1 - (\beta_0 + \sum_{j=1}^{p} \beta_j X_{1j}))^2}{\sigma^2}} \\
& \times \frac{1}{\sqrt{2\pi\sigma^2}} e^{\frac{-1}{2} \frac{(y_2 - (\beta_0 + \sum_{j=1}^{p} \beta_j X_{2j}))^2}{\sigma^2}} \ldots \\
& \times \frac{1}{\sqrt{2\pi\sigma^2}} e^{\frac{-1}{2} \frac{(y_n - (\beta_0 + \sum_{j=1}^{p} \beta_j X_{nj}))^2}{\sigma^2}}.
\end{aligned}
$$

This can be written in shorthand as

$$\prod_{i=1}^{n} \frac{1}{\sqrt{2\pi\sigma^2}} e^{-\frac{1}{2} \frac{(y_i - (\beta_0 + \sum_{j=1}^{p} \beta_j X_{ij}))^2}{\sigma^2}},$$

where $\prod_{i=1}^{n}$ indicates that the n probability densities are multiplied together. Because the constant $\frac{1}{\sqrt{2\pi\sigma^2}}$ is multiplied by itself n times, one factor in the joint probability density is

$$\left(\frac{1}{\sqrt{2\pi\sigma^2}} \right)^n = \frac{1}{(2\pi\sigma^2)^{n/2}},$$

where the exponent $1/2$ signifies the square root operation. The other factor involves the product of the n exponentials,

$$e^{\frac{-1}{2} \frac{(y_i - (\beta_0 + \sum_{j=1}^{p} \beta_j X_{ij}))^2}{\sigma^2}},$$

which, because the exponents can be added, is

$$e^{\frac{-1}{2} \sum_{i=1}^{n} \frac{(y_i - (\beta_0 + \sum_{j=1}^{p} \beta_j X_{ij}))^2}{\sigma^2}}.$$

Thus,

$$f(y_1, y_2, \ldots, y_n | \boldsymbol{\beta}, \sigma^2) = \frac{1}{(2\pi\sigma^2)^{n/2}} e^{\frac{-1}{2} \sum_{i=1}^{n} \frac{(y_i - (\beta_0 + \sum_{j=1}^{p} \beta_j X_{ij}))^2}{\sigma^2}}.$$

In matrix form, this can be written as

$$f(\mathbf{y}|\boldsymbol{\beta},\sigma^2) = \frac{1}{(2\pi\sigma^2)^{n/2}} e^{-\frac{(\mathbf{y}-X\boldsymbol{\beta})'(\mathbf{y}-X\boldsymbol{\beta})}{2\sigma^2}},$$

where $\mathbf{y}$ is the vector of observation on the dependent variable, and X is the $n \times (p+1)$ matrix of observations on the p independent variables with a leading column vector of 1's corresponding to the intercept parameter β_0, the first element in the column vector $\boldsymbol{\beta}$.

The joint probability function $f(\mathbf{y}|\boldsymbol{\beta},\sigma^2)$ is a function of the random variables $(y_1, y_2, \ldots, y_n)' = \mathbf{y}$ given the parameters $(\beta_0, \beta_1, \ldots, \beta_p)' = \boldsymbol{\beta}$. To make inferences about $\boldsymbol{\beta}$, we take the $\mathbf{y}$ values generated from the sample as fixed and regard $f(\mathbf{y}|\boldsymbol{\beta},\sigma^2)$ as a function of $\boldsymbol{\beta}$. This is denoted as $L(\boldsymbol{\beta},\sigma^2|\mathbf{y})$ rather than $f(\boldsymbol{\beta},\sigma^2|\mathbf{y})$. We call this the likelihood function. Note that $f(\mathbf{y}|\boldsymbol{\beta},\sigma^2) = L(\boldsymbol{\beta},\sigma^2|\mathbf{y})$.

We can estimate the regression parameters, $\boldsymbol{\beta}$, denoted by $\hat{\boldsymbol{\beta}}$, that maximize the likelihood function, $L(\boldsymbol{\beta},\sigma^2|\mathbf{y})$. These estimates are called maximum likelihood estimates (*MLE*). They are the estimates of the regression parameters that most likely generated the sample observations $\mathbf{y} = (y_1, y_2, \ldots, y_n)'$.

Because there is an exponential term in the likelihood function, it is easier to work with the logarithm of the likelihood function. Because the logarithmic function is a monotonic function, the value of $\boldsymbol{\beta}$ that maximizes $L(\boldsymbol{\beta},\sigma^2|\mathbf{y})$ will be identical to the value of $\boldsymbol{\beta}$ that maximizes $\log_e L(\boldsymbol{\beta},\sigma^2|\mathbf{y})$. Taking the log of both sides of $L(\boldsymbol{\beta},\sigma^2|\mathbf{y})$, we have

$$\log_e L(\boldsymbol{\beta},\sigma^2|\mathbf{y}) = -\frac{n}{2}\log_e(2\pi\sigma^2) - \frac{1}{2}\sum_{i=1}^{n}\frac{(y_i - (\beta_0 + \sum_{j=1}^{p}\beta_j X_{ij}))^2}{\sigma^2},$$

or, in matrix form,

$$\log_e L(\boldsymbol{\beta},\sigma^2|\mathbf{y}) = -\frac{n}{2}\log_e(2\pi\sigma^2) - \frac{(\mathbf{y}-X\boldsymbol{\beta})'(\mathbf{y}-X\boldsymbol{\beta})}{2\sigma^2}.$$

This is called the log-likelihood function and is generally denoted by ℓ. Because ℓ considers the y_i observations as fixed and the parameters as variables, we can write $\ell(\beta_0, \beta_1, \ldots, \beta_p, \sigma^2)$ to signify that ℓ is a multivariate function of the regression parameters.

We want to find the maximum likelihood parameter estimates of $\beta_0, \beta_1, \ldots, \beta_p$ denoted by $\hat{\beta}_0, \hat{\beta}_1, \ldots, \hat{\beta}_p$ that maximize $\ell(\beta_0, \beta_1, \ldots, \beta_p, \sigma^2)$. We can see from the likelihood that minimizing $\sum_{i=1}^{n}(y_i - (\beta_0 + \sum_{j=1}^{p}\beta_j X_{ij}))^2$ with respect to the β_j will maximize $\ell(\beta_0, \beta_1, \ldots, \beta_p, \sigma^2)$. To find the maximum likelihood estimates, $\hat{\beta}_0, \hat{\beta}_1, \ldots, \hat{\beta}_p$, we must partially differentiate the

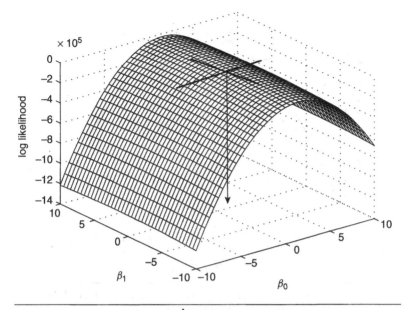

Figure 5.1 Response Surface of Log Likelihood for β_0 and β_1

log-likelihood ℓ with respect to each of the parameters, set the partial derivatives to 0, and solve the system of $(p+1)$ simultaneous equation for $\hat{\beta}_0, \hat{\beta}_1, \ldots \hat{\beta}_p$. In the case of normally distributed dependent variables, these estimating equations are identical to those resulting from employing the least squares estimating equations. This is not true for other generalized linear models, however.

Partially differentiating $\ell(\beta_0, \beta_1, \ldots, \beta_p, \sigma^2)$ with respect to $(\beta_0, \beta_1, \ldots, \beta_p)$, denoted as $\frac{\partial \ell(\beta)}{\partial \beta}$, and setting the partial derivatives to 0 results in $(X'X)\beta - X'y = 0$. Thus, $\hat{\beta} = (X'X)^{-1}X'y$, which is basically a linear transformation of y. Figure 5.1 plots the log-likelihood surface as a function of the value of β_0 and β_1 from a simple linear regression model, $y = \beta_0 + \beta_1 X_1 + e$. The maximum likelihood estimates for β_0 and β_1 are the values of β_0 and β_1 that give the maximum of log-likelihood function (the intersection of the two tangent lines). Figure 5.2 shows the one-dimensional plot of the log-likelihood function for β_0 by fixing β_1 at its maximum likelihood value $\hat{\beta}_{1(ML)}$. Notice the tangent at which the log-likelihood function reaches maximum is a horizontal line with zero gradient, and the corresponding value for β_0 is its maximum likelihood estimate. Figure 5.3 shows the one-dimensional plot of the log-likelihood function for β_1 by fixing β_0 at its maximum likelihood value $\hat{\beta}_{0(ML)}$. Again, the tangent at

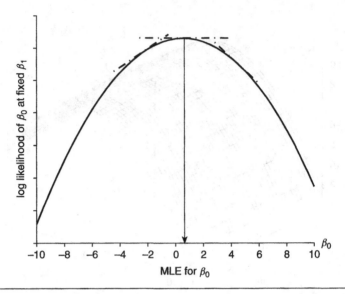

Figure 5.2   One-Dimensional Log Likelihood for β_0 With Value of β_1 Fixed to Its Maximum Likelihood Estimate

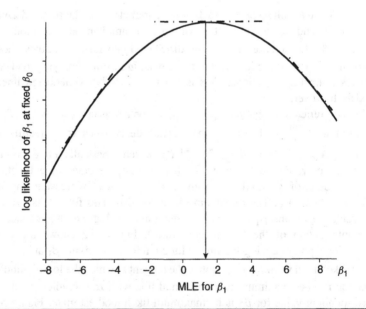

Figure 5.3   One-Dimensional Log Likelihood for β_1 With Value of β_0 Fixed to Its Maximum Likelihood Estimate

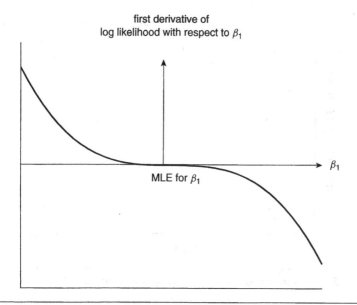

Figure 5.4 First Derivative of Log Likelihood in the Neighborhood of *MLE* for β_1 (at fixed β_0)

which the log-likelihood function reaches its maximum has zero gradient. In general, the gradient of the tangent line at any value of β_1 can be determined by computing the first derivative of the log-likelihood function.

Figure 5.4 shows the value of the first derivative (gradient of the tangent) in the neighborhood of $\hat{\beta}_1$. Notice the derivative is zero at $\hat{\beta}_{1(ML)}$ but is nonzero at other values of β_1.

Let us examine two different likelihood functions and determine what they tell us about the precision of our estimate of $\hat{\beta}_{1(ML)}$. Figure 5.5 shows a likelihood function (based on a data set generated from a sample of 10 subjects from a regression model) that is rather flat and wide around a large neighborhood of $\hat{\beta}_{1(ML)}$. In other words, a wide range of $\hat{\beta}_1$'s give approximately the same likelihood value, and we cannot precisely determine the best estimate for β_1. The gradients of the tangent line in this range of $\hat{\beta}_1$'s are all close to zero. On the contrary, the likelihood function (based on a data set generated from a sample of 1,000 subjects from the same regression model used previously) is much "sharper" in Figure 5.6, and the maximum likelihood estimate for β_1 can then be determined more precisely. As can be seen from the figure, the gradient of the tangent line around the maximum likelihood estimate for β_1, $\hat{\beta}_{1(ML)}$, changes abruptly after the value β_1 moves away from $\hat{\beta}_{1(ML)}$.

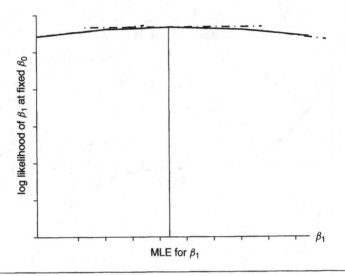

Figure 5.5   Likelihood Function for an Imprecise Estimate of β_1

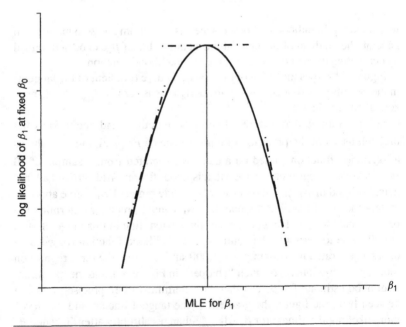

Figure 5.6   Likelihood Function for a Precise Estimate of β_1

Generalized linear models other than normal regression result in a system of equations for estimating regression parameters that cannot be solved analytically. We must resort to various iterative numerical algorithms that eventually converge to the maximum likelihood estimates through repeated cycles of the algorithm.

6. DEVIANCE AND GOODNESS OF FIT

The goal of modeling is to find a set of independent variables that results in a model for μ_i that provides a good fit to the observed values y_i. For normally distributed dependent variables, one goodness of fit criterion is $\sum_{i=1}^{n}(y_i - \hat{\mu}_i)^2$. The smaller $\sum_{i=1}^{n}(y_i - \hat{\mu}_i)^2$, the better the model $\hat{\mu}_i = \beta_0 + \sum_{j=1}^{p}\beta_j X_{ij}$ fits the data. This is the error sum of squares, which is minimized in either least squares or maximum likelihood estimation of the regression parameters β_j. The error sum of squares when divided by the population variance, σ^2, is also called the deviance, which is a term used for generalized linear models to describe how well the model fits the data based on a statistical criterion that measures goodness of fit.

The deviances associated with other generalized linear models, such as the Poisson and logistic regression models, are different from the normal case and from each other because the data generated from these models have their own unique probability function or likelihood functions. The deviance can also be used to compare the fit of two models by taking the differences in the deviances. For example, if the deviance was not much smaller for a model containing more regression parameters as contrasted to a model containing fewer regression parameters, then we may choose the model with few regression parameters because of its parsimony and ease of interpretation.

The difference in the deviance between the more complex, hereafter called the full model, and the deviance of the simpler model with some parameters dropped out, hereafter called the reduced model, can also be used to test the null hypothesis that the additional parameters in the full model are equal to zero. The difference in the deviances is a chi-square distributed variable (under the null hypothesis that the additional regression parameters are equal to zero) with degrees of freedom equal to the difference in the number of regression parameters between the full and reduced model. That is, for these other generalized linear models, the difference in the fit between the full and reduced model can be used directly in a statistical test because its value is distributed as chi-square under the null hypothesis that the additional regression parameters are zero. The degree of freedom is equal to the difference in the number of parameters. The normal model, however, has a nuisance parameter, the unknown population variance, σ^2, in its deviance. If σ^2 is unknown and has to be estimated from the samples, we cannot use the chi-square distribution but have to use the F distribution.

Let us now formally define deviance. Deviance (D) is defined as $2[\ell(b_{max}|y) - \ell(b|y)]$, where $\ell(b_{max}|y)$ is the log likelihood when the estimated conditional mean $\hat{\mu}_i$ for each observation is set to its observed value y_i. This is the model that yields the largest log likelihood because there is a separate parameter $\hat{\mu}_i = y_i$ for each sample member. There are as many parameters as sample members; therefore, the model fits the data perfectly. Thus, we have not simplified things using a model like this. It does, however, indicate the largest likelihood possible with the observed sample of y_i's. It gives us a benchmark for evaluating other less complex models. The second term in the brackets, $\ell(b|y)$, is the log likelihood for the simpler model whose parameters $\beta = (\beta_0, \beta_1, \ldots, \beta_p)'$ are estimated by maximum likelihood and are used to generate the $\hat{\mu}_i$'s. Two times the difference in the log likelihood gives us an indication of how well the model under investigation fits the data.

For example, if we had 100 observations on y_i, a particular normally distributed dependent variable, along with three independent variables, X_{i1}, X_{i2}, and X_{i3}, then $\ell(b_{max}|y)$ would be calculated by substituting y_i for $\hat{\mu}_i$ in the log likelihood, whereas $\ell(b|y)$ would be calculated by estimating $\hat{\beta}_0$, $\hat{\beta}_1$, $\hat{\beta}_2$, and $\hat{\beta}_3$ and substituting $\hat{\mu}_i = \hat{\beta}_0 + \hat{\beta}_1 X_{i1} + \hat{\beta}_2 X_{i2} + \hat{\beta}_3 X_{i3}$ for each of the 100 estimated $\hat{\mu}_i$'s in the log-likelihood equation. The maximal model contains 100 parameters, whereas the model under consideration contains only 4 parameters. The deviance, D, indicates how well the simpler model fits the data compared to the maximal model, which fits the data exactly.

To compare the two models, we compute the deviance for each of the two models and subtract the deviance of the complex model or full model (D_F) from the deviance of the simpler or reduced model (D_R). That is, we calculate $D_R - D_F$. If this difference is large, then there is evidence that the full or more complex model provides a better fit to the data. The deviance for the reduced or simple model will always be larger than the deviance for the full or more complex model. The question is whether it is large enough to warrant including the extra regression parameters in the model.

Using Deviances to Test Statistical Hypotheses

We can formally test the hypothesis that the additional parameters in the full model are zero by using the difference in deviances in a formal statistical test. Under the null hypothesis that the additional parameters are zero, $(D_R - D_F)$ is distributed as chi-square with degrees of freedom equal to the difference in the number of parameters in the two models or, equivalently, the number of additional parameters in the full model. For generalized linear models, such as the Poisson and logistic models, we can calculate a numerical value for $(D_R - D_F)$ and refer the value to a chi-square table with the appropriate degrees of freedom, the difference in the number of parameters, to determine if the chi-square is significant.

Goodness of Fit

We can compare the fits of various alternative models by contrasting their associated deviances. We can also compare the fit of various link functions. In addition, we should examine the residuals of the model to determine if there are some observations that have large residuals, which indicate a poor fit of the model to the independent variable for those observations.

If the deviance of the final selected model is large and statistically significant when referred to a chi-square distribution with the appropriate degrees of freedom (sample size minus the number of parameters), then this may indicate overdispersion. Overdispersion means that the estimated variance of the dependent variable about its conditional mean is larger than would be expected for the probability distribution used in the model. For example, the Poisson distribution assumes that the variance of the distribution is equal to its mean. If the data suggest that it is larger, then the standard error of the regression parameters may have to be adjusted upwards, although the parameter estimates are unbiased.

Assessing Goodness of Fit by Residual Analysis

Like the regression model based on the normal distribution, the analysis of various kinds of residuals appropriate to generalized linear models can be used to help identify poorly fitting observations that are not well explained by the model. Systematic departures of the values of the dependent variables from the predicted values based on the model can be checked from residual plots. The analysis of residuals can help in respecifying a model that provides a significantly better fit to the data. For example, it may provide evidence of the need for a quadratic term in the model, such as the square of one or more of the independent variables, or it may provide evidence that the observations on the dependent variable are correlated and thus violate the independence assumptions of the likelihood function. There are generalized linear models that can accommodate correlated observations in which the correlations are due to repeated outcome measures on the same individuals or the clustering of individuals within homogeneous groups such as classes within schools. These more complex models, however, are not discussed here.

Probably the simplest residual to assess goodness of fit for generalized linear models is the Pearson residual, which is

$$r_i = \frac{y_i - \hat{\mu}_i}{\sqrt{\text{var}(\hat{\mu}_i)}},$$

where the hats indicate that the mean and variance of an observation are estimated from the model. For example, in the case of a logistic regression model, $y_i = 1$ or 0,

$$\hat{\mu}_i = \hat{\pi}_i = \frac{e^{\sum_{j=0}^{p} \hat{\beta}_j X_{ij}}}{1 + e^{\sum_{j=0}^{p} \hat{\beta}_j X_{ij}}},$$

and $\text{var}(\hat{\mu}_i) = \hat{\pi}_i(1 - \hat{\pi}_i)$. Note that unlike the normal model, in which the variance, σ^2, is assumed to be constant across observations, each observation, y_i, in the logistic regression model has a unique variance based on the regression parameters and the corresponding values of the independent variables, X_i. The Pearson residual for a logistic regression model is

$$r_i = \frac{y_i - \hat{\pi}_i}{\sqrt{\hat{\pi}_i(1 - \hat{\pi}_i)}}.$$

Large values of r_i indicate a lack of fit for that observation.

In the case of the Poisson regression model,

$$r_i = \frac{y_i - \hat{\mu}_i}{\sqrt{\text{var}(\hat{\mu}_i)}} = \frac{y_i - \hat{\mu}_i}{\sqrt{\hat{\mu}_i}},$$

where

$$\hat{\mu}_i = \hat{\lambda}_i = e^{\sum_{j=0}^{p} \hat{\beta}_j X_{ij}}.$$

For the Poisson distribution, the variance is equal to its mean, and thus $\hat{\mu}_i = \text{var}(\hat{\mu}_i)$ in the previous equation. Again, the variance of the outcome variable, y_i, which for the Poisson model is a count, varies with the predicted mean for that observation.

Another residual that is commonly used is the deviance residual. The deviance is composed of the sum of the deviances of the individual observations. The contribution to the deviance from a single observation, y_i, is a measure of how well the model fits a particular observation y_i. Like the

Pearson residual, its definition depends on the form of the deviance associated with a particular generalized linear model.

The deviance residual is

$$r_i = sign(y_i - \hat{\mu}_i) \sqrt{2(\ell_i(y_i) - \ell_i(\hat{\mu}_i))}.$$

The term $sign$ $(y_i - \hat{\mu}_i)$ indicates whether the residual $(y_i - \hat{\mu}_i)$ is positive or negative. The $\ell(y_i)$ term is the value of the log likelihood when the mean of the conditional distribution for the ith individual is the individual's actual score of the dependent variable; $\ell(\hat{\mu}_i)$ is the log likelihood when the conditional mean, generated by the model, is substituted in the log likelihood. The term under the square root sign is the contribution of the ith observation to the total deviance, which, as noted earlier, is equal to $2 \sum_{i=1}^{n} (\ell(y_i) - \ell(\hat{\mu}_i))$. For example, the deviance residual for the Poisson distribution is

$$r_i = sign(y_i - \hat{\mu}_i) \sqrt{2[y_i \log(y_i/\hat{\mu}_i) - (y_i - \hat{\mu}_i)]}.$$

7. LOGISTIC REGRESSION

We briefly introduced the logistic regression in Chapter 4. Using some of the modeling concepts discussed previously, we now show how logistic regression fits into the generalized linear model framework. An example of applying logistic regression to a real data set is also presented. The Bernoulli distribution, which is a special case of the binomial distribution, is used to model binary (0,1) outcomes or dependent variables, such as treatment success (1) versus treatment failure (0) or smoking relapse (1) versus no relapse (0). Which of the events are coded 1 or 0 is completely arbitrary, although, in general, the event of interest is coded 1. Other examples of binary outcome variables are dead versus alive and crime committed versus no crime committed. Many other outcomes in the social sciences and other sciences are binary. They are the dependent variables for the logistic regression model.

The Bernoulli or binary distribution is $f(y|\pi) = \pi^y(1 - \pi)^{1-y}$, where π is the probability of a successful outcome ($y = 1$). There are two outcomes (values of y): $y = 1$ if the outcome is a success, and $y = 0$ if the outcome is a failure. Note that if we substitute $y = 1$ in the probability distribution, we get $f(1|\pi) = \pi$. If we substitute $y = 0$, we get $f(0|\pi) = 1 - \pi$. If there are only two outcomes, and one occurs with probability π, then the other outcome must occur with probability $(1 - \pi)$ because the sum of the probabilities must equal 1.

This is a rather simple probability distribution from the exponential family that can be easily put into exponential form. We have

$$f(y|\pi) = e^{y \log_e(\frac{\pi}{1-\pi}) + \log_e(1-\pi)} = e^{y\theta - \log_e(1+e^\theta)},$$

where the canonical parameter $\theta = \log_e \frac{\pi}{1-\pi}$, and $b(\theta) = \log_e(1 + e^\theta)$. In this case, the dispersion parameter $\phi = 1$, and $c(y, \phi) = 0$. The canonical link function is $\theta = \log_e \frac{\pi}{1-\pi}$, and θ is the parameter that we model as a linear function of the covariates—that is, $\theta = \beta_0 + \sum_{j=1}^p \beta_j X_j$. The variance function is $b''(\theta) = \pi(1 - \pi)$. It is a one-parameter distribution, and its variance, $\pi(1 - \pi)$, is related to its mean π. The log likelihood for the Bernoulli distribution is $\ell(\pi|y) = y \log_e \pi + (1 - y)\log_e(1 - \pi)$. The log likelihood for the whole sample is $\ell(\pi_1, \ldots, \pi_n|y_1, \ldots, y_n) = \sum_{i=1}^n [y_i \log_e \pi_i + (1 - y_i)\log_e(1 - \pi_i)]$.

As before, we assume that each observation or sample member comes from a Bernoulli distribution with its own unique parameter π_i. We want to write the log likelihood $\ell(\pi|y)$ as a function of the regression parameters (i.e., $\ell(\beta|y)$) because they, and not π_i, are what we want to estimate from the data. Some algebra shows that

$$\pi_i = \frac{e^{\sum_{j=0}^p \beta_j X_{ij}}}{1 + e^{\sum_{j=0}^p \beta_j X_{ij}}},$$

and this is what we substitute into the log likelihood.

To find the maximum likelihood estimates of the parameters, we substitute $\pi_i(\beta)$ as a function of the regression parameters into the log-likelihood equation, partially differentiate it with respect to each of the parameters, set the partial derivative equations to 0, and solve for the vector of regression parameters, β. Because the equations are nonlinear in the parameters and cannot be solved by analytic methods, an iterative algorithm such as iterative reweighted least squares are needed to solve them for β.

The deviance for the logistic model is $D = 2[\ell(y|y) - \ell(\hat{\beta}|y)]$, where y is the vector of binary outcome variables $(y_1, y_2, \ldots, y_n)'$, and $\hat{\beta}$ is the vector of maximum likelihood regression parameter estimates. The likelihood for the saturated model is $\ell(y|y)$, where the maximum likelihood estimate of the parameter π_i is y_i; that is, there is a unique parameter estimate for each observation. For the Bernoulli distribution, $\ell(y|y) = \sum_{i=1}^n [y_i \log_e y_i + (1 - y_i)\log_e(1 - y_i)]$. Note that when y_i is either 0 or 1, the term within the brackets is equal to 0 so that $\ell(y|y) = 0$. Thus, the deviance simplifies to

$$D = -2\ell(\hat{\beta}|y) = -2\sum_{i=1}^n [y_i \log_e \hat{\pi}_i(\hat{\beta}) + (1 - y_i)\log_e(1 - \hat{\pi}_i(\hat{\beta}))],$$

where $\hat{\pi}_i(\hat{\boldsymbol{\beta}})$ indicates that the maximum likelihood estimate, $\hat{\pi}_i$, is a function of the maximum likelihood regression parameter estimates, $\hat{\boldsymbol{\beta}} = (\hat{\beta}_0, \hat{\beta}_1, \ldots, \hat{\beta}_p)'$. As mentioned previously,

$$\hat{\pi}_i(\hat{\boldsymbol{\beta}}) = \hat{p}_i = \frac{e^{\sum_{j=0}^{p} \hat{\beta}_j X_{ij}}}{1 + e^{\sum_{j=0}^{p} \hat{\beta}_j X_{ij}}}.$$

Thus, for a particular logistic regression model, $D = -2\ell(\hat{\boldsymbol{\beta}}|y)$ measures how well the model under consideration fits the data. As mentioned previously, the difference in the deviance for two different regression models, where one model is a reduced model that contains a subset of the parameters in the full model, is distributed as chi-square with degrees of freedom equal to the difference in the number of regression parameters in the full and reduced model. It compares the fit of a particular model with the fit of a simpler model with fewer parameters. If $(D_R - D_F)$ following chi-square distribution is not significant for the appropriate degrees of freedom, then there is evidence that the extra parameters in the full model are not necessary, and that a simpler model with fewer regression parameters is sufficient.

If we want to test whether a particular single regression parameter is statistically significant, then we can form the t ratio of $\frac{\hat{\beta}_j}{\hat{\sigma}_{\beta_j}}$ and compare the t ratio to a t distribution with $(n - p - 1)$ degrees of freedom to determine if it is statistically significant. The denominator is the standard error of the regression parameter estimate. We could also test the significance of a regression parameter by taking the difference in deviances with and without the parameter in the model and refer to a chi-square table with one degree of freedom.

Like all regression models, major interest focuses on the parameter estimates, their estimated standard errors, their t ratios, and their statistical significance levels. This information forms the core output of all software packages that estimate generalized linear models. For each type of generalized linear model, the regression parameters have a different interpretation with respect to their effect on the dependent variable.

For logistic regression, the link function, which is a nonlinear function of the mean, that is, $\log_e(\frac{\pi_i}{1 - \pi_i})$, is a linear function of the regression parameters. Thus, β_j indicates the change in $\log_e(\frac{\pi_i}{1 - \pi_i})$ due to a one-unit increase in X_{ij} rather than indicating the change in π_i, the mean of the Bernoulli distribution.

Interpreting β_j is difficult because it reflects a change in $\log_e(\frac{\pi_i}{1 - \pi_i})$. If we exponentiate β_j (i.e., e^{β_j}), then e^{β_j} measures the odds ratio associated with a one-unit increase in X_i. Let us take a simple example of a logistic

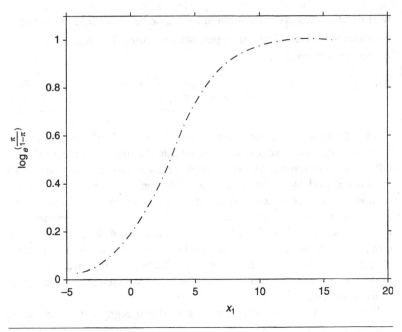

Figure 7.1   Logistic Regression Function

regression with a single independent variable to illustrate this. Let $\log_e(\frac{\pi}{1-\pi}) = \beta_0 + \beta_1 X_1$. Figure 7.1 shows this function.

If we exponentiate both sides of this logistic regression equation, then

$$\frac{\pi}{1-\pi} = e^{\beta_0 + \beta_1 X_1} = e^{\beta_0} e^{\beta_1 X_1}.$$

If we increase X_1 by one unit, then

$$\frac{\pi^*}{1-\pi^*} = e^{\beta_0 + \beta_1 (X_1 + 1)} = e^{\beta_0} e^{\beta_1 X_1} e^{\beta_1}.$$

We can find e^{β_1} by dividing the second equation by the first equation. That is,

$$\frac{\frac{\pi^*}{1-\pi^*}}{\frac{\pi}{1-\pi}} = \frac{e^{\beta_0} e^{\beta_1 X_1} e^{\beta_1}}{e^{\beta_0} e^{\beta_1 X_1}} = e^{\beta_1}.$$

Thus, e^{β_1} indicates an increase or decrease in the odds due to a one-unit increase in X_1. If β_1 is positive, then e^{β_1} is greater than 1, which indicates an increase in the odds due to a one-unit increase in X_1. For example, if β_1 equals .1, then $e^{0.1}$ equals 1.11, and there is an increase in the odds of 11% due to a one-unit increase in X_1. A large positive value of β_1 results in a large value of e^{β_1}, which in turn indicates a large increase in the odds due to a one-unit increase in X_1. If β_1 is negative, however, then e^{β_1} is less than 1, which indicates a decrease in the odds due to a one-unit increase in β_1. For example, if β_1 equals $-.1$, then $e^{-0.1}$ equals 0.90, and there is a decrease in the odds of 10% due to a one-unit increase in X_1. A large negative value of β_1 results in a small value of e^{β_1} and a large decrease in the odds.

There are other link functions (e.g., probit) for modeling binary variables besides the logistic link, but they give essentially the same results as the logistic regression. Because the parameters of the logistic are easier to interpret than those for other models, it is the most widely used regression model for binary dependent variables.

Example of Logistic Regression

The first author was involved in a number of state-level studies designed to estimate drug use prevalence and treatment needs using social indicators. The underlying premise of the social indicator approach at the state level is that various social, demographic, and economic characteristics of counties or other local planning entities within a state are associated with drug use prevalence and the need for treatment. Examples of social indicators available at the county level through various county, state, and federal agencies include median household income; demographic distributions by age, gender, and race; rates of alcohol- or drug-related traffic accidents; and violent crime rates.

Although surveying the state population directly on the use of drugs and alcohol is probably the best approach for obtaining information, this approach has some serious limitations. Surveys are expensive to implement year after year to gauge the drug use problem. Even with relatively large sample sizes of 4,000 or 5,000 respondents at the state level, the sample sizes will be too small to make any county-level drug use estimates. It is not unusual for a state to have more than 50 or even 100 counties so that the average county-level sample size would range from 50 to 100. These sample sizes are much too small to make any county-level inferences concerning drug use and treatment needs.

If we can isolate county-level social indicator variables that are predictive of various county-level drug use and treatment measures based on state-level telephone survey data, then we can achieve two major objectives. First, we can use the logistic regression model to make county-level model-based estimates that are more precise than sample survey estimates based on extremely small county-level samples. That is, the direct survey estimates of drug use prevalence measures at the county level are based on 50 to 100 respondents on average, whereas the indirect logistic regression model-based estimates, for example, use the data for all the 4,000 or 5,000 survey respondents in the state.

Once the model is estimated, then a county-level prevalence measure for a particular drug or type of treatment need can be estimated by substituting the values of the social indicators for a particular county into the logistic regression equation to obtain the probability of drug use or treatment need in that county. In this situation, we are not interested in estimating the odds of drug use but, rather, the probability of drug use. Because

$$\frac{\pi}{1 - \pi} = e^{\sum_{j=0}^{p} \beta_j X_j},$$

a little algebra shows that

$$\pi = \frac{e^{\sum_{j=0}^{p} \beta_j X_j}}{1 + e^{\sum_{j=0}^{p} \beta_j X_j}} = \frac{1}{1 + e^{-\sum_{j=0}^{p} \beta_j X_j}}.$$

Thus, once the β_j's are estimated, we can substitute the X_j's of the associated social indicators for a particular county into the previous formula to obtain the predicted drug use for that county.

As well, we can use this model to predict future drug use when expensive drug use survey data are not available. Many of the county-level social indicator variables are dynamic variables whose values change over time (e.g., various crime rates). Thus, social indicator data can be collected in future years and substituted into the previous equation to predict drug abuse in future years. In most cases, only a handful of social indicator variables are important predictors of drug use prevalence, so only a small amount of information needs to be gathered to substitute into the previous equation. The logistic regression equation or, equivalently, the previous formula captures the relationship between the county-level social indicators and county-level probability of drug use. As the social indicators change through time, there are corresponding changes in the predicted drug use prevalence.

The logistic regression model presented here is based on a survey study conducted in South Dakota. The telephone survey of drug use and treatment need used to calibrate the logistic regression social indicator model (i.e., the dependent binary drug use variables at the respondent level) yielded a total sample of 4,205 respondents. There are 66 counties in South Dakota, and the county-level sample sizes ranged from 4 to 896. The county-level sample sizes were approximately proportional to the population sizes of the counties, as would be expected. Most of the county sample sizes were 50 or less. A number of drug use prevalence measures were modeled, including past year heavy drinking, past year illicit drug use, need for alcohol treatment, and need for drug treatment. Data on 41 county-level social indicator variables were collected from various sources. They included indicators of alcohol and drug use (e.g., adult arrest rate for drug use or possession), community disorganization (e.g., divorce rate), community crime and violence (e.g., adult arrest rate for violent crimes), demographic characteristics (e.g., percentage of county population that is White), socioeconomic deprivation (e.g., unemployment rate), alcohol and drug availability (e.g., distance to nearest interstate highway), academic failure and lack of commitment (e.g., high school drop-out rate), and problems indirectly associated with substance abuse (e.g., teenage birth rates).

Because there were so many indicators and many were highly correlated, factor analysis was used to cluster the variables, and a few indicators were chosen to represent each cluster of variables or, equivalently, to measure each factor. This reduced the initial set of social indicators. The reduced set of social indicators was still too large to use its independent variables simultaneously in the logistic regression models predicting the various drug use measures, so stepwise logistic regression was used to build more parsimonious models.

We use one of these models for our example. The dependent variable was the binary variable, need for some kind of intervention or treatment for either alcohol use or drug use. It was coded 1 if the respondent had an intervention need and 0 otherwise. The four county-level social indicators or independent variables were juvenile liquor law violations ($JLLV$), measured in number of juvenile liquor law violations per 1,000 juveniles; juvenile arrest rate for violent crime (JVC), measured in number of arrests per 1,000 juveniles; percentage of young males (YM), measured as a percentage of the county population; and median income level (MI) for the county, measured in dollars. The results of the logistic regression analyses are presented in Table 7.1.

The logistic regression results in Table 7.1 are a modification of the results of a SAS run. Note that the estimated regression parameters for $JLLV$ and MI are extremely small. They had odds ratios of 0.993 and 0.999, respectively. This is partly due to the measurement scale of the associated independent variables. The odds ratio for juvenile liquor law violations indicates that an increase of one liquor law violation per 1,000 juveniles decreases the odds of drug treatment intervention need by 0.7% (100 $(1 - 0.993)$%), whereas a $1 increase in median income decreases the intervention need by 0.1% (100 $(1 - 0.999)$%).

The standard deviation for juvenile liquor law violations was 14.1 violations per 1,000 juveniles so that a 1-violation increase per 1,000 is a comparatively minor change. If we change the measurement scale of $JLLV$ to standard deviation units, then a standard deviation increase in $JLLV$ results in an odds ratio of $(0.993)^{14.1}$ or 0.906. Thus, a standard deviation

TABLE 7.1
Logistic Regression Model for Drug Use Intervention

Variable	Parameter Estimate	Standard Error	t Ratio	Significance Level	Odds Ratio
Intercept	3.7757	.4202	8.985	<.0001	—
JLLV	−0.00707	.00317	−2.230	.0321	0.993
JVC	0.1964	.0586	3.352	.0019	1.217
YM	−0.0471	.0153	−3.078	.0040	0.954
MI	−0.00007	.000016	−4.375	.0001	0.999

increase in *JLLV* decreases the odds of intervention need by 9.4%. Even in standard deviation units, this is only a moderate effect for *JLLV*.

A more meaningful change in median income may be measured in $1,000 units rather than $1 units. Median income ranged from $11,502 to $34,286 across the 66 South Dakota counties. If we change the median income scale to $1,000 units, then the associated odds ratio becomes 0.368. This is a large effect because an increase of $1,000 in median income decreases the odds of intervention need by 63.2%.

The juvenile arrest rate for violent crimes per 1,000 juveniles ranged only from 0 to 3.73, with many counties registering no arrests so that the original metric of 1 arrest per 1,000 seems to be a reasonable metric. The odds ratio associated with *JVC* was 1.217, which indicates that an increase of a single arrest per 1,000 increases the odds of intervention need by 21.7%, a moderate effect. The percentage of young males (15-34 years old) in the county population ranged from 9% to 23%, with a standard deviation of 2.7%. The effect of a standard deviation increase in the percentage of young males in the county on the odds of intervention need is $(0.954)^{2.7} = 0.881$. That is, a standard deviation increase in the percentage of young males decreases the odds of an intervention need by 11.9%, a moderate effect.

The positive effect of juvenile violent crimes and the negative effect of median income on odds of intervention need are expected. The negative effect of percentage of young males and the negative effect of juvenile liquor law violation rate are somewhat counterintuitive. Perhaps, high rates of liquor law violations signify high diligence on the part of law enforcement authorities to keep underage drinking in check, thereby reducing the need for intervention. It is more difficult to explain the negative effect of percentage of young males. One needs to keep in mind, however, that the effects of each of the four social indicators are adjusted for the effects of the remaining three variables. The pattern of interrelationships among the variables in the model may result in adjusted effects (e.g., % males) being different in direction than unadjusted effects based on bivariate models.

To obtain the predicted intervention need for a county, we simply substitute the values of the corresponding four social indicators for a particular county into the following formula:

$$\hat{p}_{\text{need}} = \frac{1}{1 + e^{-(3.7757 - 0.00707\,JLLV + 0.1964\,JVC - 0.0471\,YM - 0.00007\,MI)}}.$$

8. POISSON REGRESSION

The Poisson regression model assumes that the random component of the regression model has a specific probability distribution that in this case is the Poisson distribution. The Poisson distribution is applicable to count data. By count data, we mean the number of times a particular event occurs in a given period of time. The following are examples of count data in which the Poisson distribution is applicable: the number of traffic accidents at a given busy intersection during a given period of time, such as a year; the number of telephone calls a switchboard receives in an hour; the number of crimes committed by a criminal during 1 year; the number of treatment episodes of a drug addict during 5 years; and the number of emergency room admissions for a drug overdose at a particular hospital during a given period of time.

The Poisson probability density function is a much simpler looking expression than the normal probability density function. It is

$$f(y|\lambda) = \frac{\lambda^y e^{-\lambda}}{y!},$$

where $e = 2.7183$ (the base of the natural logarithm), and $y! = y(y-1)(y-2)\cdots 1$. For example, $6! = 6 \times 5 \times 4 \times 3 \times 2 \times 1$. There is only one parameter, θ, in the Poisson distribution, which is the mean number of events in a given time period. Figure 8.1 shows the Poisson distributions at various values of λ.

The Poisson distribution is a member of the exponential family because

$$f(y|\lambda) = e^{(y \log \lambda - \lambda - \log y!)} = e^{(y\theta - e^\theta - \log y!)}.$$

Thus, the canonical parameter θ equals $\log_e \lambda$, which is also the canonical link. The variance function $b(\theta)$ is e^θ, whose second derivative $b''(\theta) = e^\theta$. Because $e^\theta = \lambda$, the variance of the Poisson distribution is equal to its mean. As the mean of a Poisson distribution increases, so does its variance. The random variable y can take on only nonnegative integer values—that is, $0, 1, 2, \ldots$. The Poisson distribution is skewed to the right.

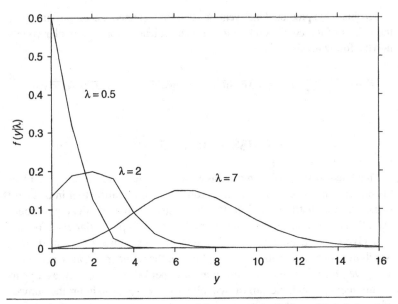

Figure 8.1   Poisson Distribution for Different Parameter Values of λ (0.5, 2, and 7)

For the Poisson regression model, assuming a canonical link, we assume that $\log_e \lambda$ is a function of a linear model. That is, $\log_e \lambda = \sum_{j=0}^{p} \beta_j X_j$. We substitute $\sum_{j=0}^{p} \beta_j X_j$ into the log likelihood for the sample and solve for the β_j's that maximize the likelihood function.

Because we are concerned with regression models, we can test whether a regression model with p regression parameters fits the data better than a simple model with simply an overall mean, which can be considered as a regression model with only an intercept parameter in the model. We do this as before: Compute the difference in the deviance between the reduced model with simply an overall mean or intercept parameter and the full model, where the λ_i, the conditional means, are a function of the regression parameters. We can also compare two models, a full and reduced model, by computing the difference in the deviances between the two models. The difference is distributed as chi-square with degrees of freedom equal to the difference in the number of regression parameters. If the chi-square is statistically significant, then we accept the full model. If it is not significant, we accept the reduced model.

As mentioned previously, deviance is defined as $2[\ell(y|y) - \ell(\hat{\lambda}|y)]$, where $\ell(y|y)$ is the maximum likelihood achievable in which the fitted values equal the data values. That is, there are as many parameters as observations. The second likelihood, $\ell(\hat{\lambda}|y)$, is the likelihood based on the predicted values $\hat{\lambda}_i$, which are generated by a regression model where $\hat{\lambda}_i$ is a function of $\hat{\beta}$—that is, $\hat{\lambda}_i(\hat{\beta})$. For the Poisson distribution, the deviance is

$$D = 2\left\{\sum_{i=1}^{n}\left[y_i\,\log\left(\frac{y_i}{\hat{\lambda}_i}\right) - (y_i - \hat{\lambda}_i)\right]\right\}.$$

The object of modeling is to find a model with a low value of D because this indicates that the conditional means or predicted values $\hat{\lambda}_i$ are close to the observed values y_i. If the model fits the data perfectly (i.e., as well as a saturated model), then D equals 0 because $y_i = \hat{\lambda}_i$. We can see from the formula that as the difference between y_i and $\hat{\lambda}_i$ increases, D becomes larger. Thus, D can be used as a measure of goodness of fit of the model under consideration. Because D is the sum of individual deviance terms— that is, $D = \sum_{i=1}^{n} d_i$—we can examine the individual d_i's and determine if there are any values of d_i that are extremely large. What is known as a deviance residual is defined as sign $(y_i - \hat{\lambda}_i)\sqrt{d_i}$, where sign $(y_i - \hat{\lambda}_i)$ equals positive if $y_i > \hat{\lambda}_i$, equals negative if $y_i < \hat{\lambda}_i$, and equals 0 if $y_i = \hat{\lambda}_i$. Large deviance residuals may indicate a lack of it.

For the Poisson model, because $\log_e \lambda_i = \sum_{j=0}^{p} \beta_j X_{ij}$ implies

$$\lambda_i = e^{\sum_{j=0}^{p} \beta_j X_{ij}} = e^{\beta_0} e^{\beta_1 X_{i1}} e^{\beta_2 X_{i2}} \ldots e^{\beta_p X_{ip}},$$

the effect of a one-unit increase in X_{ij} is to change the value of λ_i, the mean, by a factor of e^{β_j}, assuming that the other independent variables are held fixed.

Example of Poisson Regression Model

The Poisson regression model described here is a real-life example of its application to an important public policy issue. The first author was a collaborator in this research effort. The research involved the study of disciplinary infraction rates among inmates in the state prisons in North Carolina. It was observed by prison officials and other workers in the prison system that the disciplinary infraction rate among prisoners had been substantially increasing in recent times. There was a belief that this increase was due to a change in the sentencing laws. Prior to October 1, 1994, convicted felons were sentenced under the Fair Sentencing Act (FSA). Under FSA, inmates had relatively long sentences but could get them reduced by half and were eligible for parole. On October 1, 1994, convicted felons were sentenced under the Structured Sentencing Act (SSA). This law provides for relatively short sentences and no parole eligibility. There is no early release for good behavior. The inmate can gain some earned time for early release for participating in certain work assignments and programs. Earned time, however, cannot cause an inmate to serve less than 83% of his or her maximum sentence.

Thus, SSA inmates have less incentive to comply with prison disciplinary regulations than FSA inmates and, consequently, would be expected to have a higher disciplinary infraction rate. This would result in a number of adverse consequences for the safety of staff and inmates.

This study was designed to determine if inmates sentenced under SSA had significantly higher disciplinary infraction rates than inmates sentenced under FSA. The study involved male and female inmates of all ages who were admitted to North Carolina prisons for several years beginning June 1, 1995. Some inmates, such as those serving sentences under both FSA and SSA and those returned to prison because of parole violations, were excluded. There were a few other exclusions that need not be discussed here. There were a substantial number of both FSA and SSA inmates during the study period. The data used in the regression analyses came from the North Carolina Department of Correction's computerized inmate records.

The research involved modeling different types of disciplinary infraction rates separately for both males and females. Males and females were modeled separately because males had a much higher infraction rate, and it was suspected that the effects of FSA versus SSA and other covariates on disciplinary infractions differed between male and female inmates. That is, it was hypothesized that gender interacted with FSA versus SSA and most of the other covariates. We focused on the Poisson regression modeling of the total or overall disciplinary infraction rate for male inmates. This was probably the most important of all the Poisson regression models that were estimated in this research project because the prison population is mostly male, and their infraction rate is higher than that of female inmates.

The sample size for the Poisson regression model was 11,738. The records for these inmates had no missing values. An additional 1,026 inmates were dropped from the analysis because of missing values. Because inmates sentenced under FSA and SSA could differ in important background characteristics that would be expected to be related to disciplinary infraction rates, they needed to be included in the model as covariates along with the FSA versus SSA "treatment" or policy variable to adjust the FSA versus SSA policy variable for the effects of the covariates. There were a large number of possible covariates on the inmates' records from which to select, but we decided on a subset described later on the basis of logical considerations and past research.

The primary variable of interest was the policy variable Structured Sentencing versus Fair Sentencing, which was entered in the model as a 0-1 indicator model, with 0 indicating Fair Sentencing and 1 indicating Structured Sentencing. Thus, Fair Sentencing represents the reference group to which Structured Sentencing will be compared. The regression parameter associated with this 0-1 variable indicates the effect of Structured Sentencing relative to Fair Sentencing. The parameter is adjusted for the effects of the other covariates in the model.

There was a three-level categorical variable reflecting the type of crime for which the inmate was sentenced. The first level was violent crimes; the second level was property crimes; and the third level was public order crimes, such as drug offenses. It was hypothesized that inmates sentenced for violent crimes would have higher disciplinary infraction rates than inmates convicted of the nonviolent crimes represented by the other two levels. Because there are three categories of crimes, two 0-1 indicator variables are required to represent this three-level crime variable. Note that a third indicator is redundant because if the values of any two indicator variables are known, then the value of the remaining indicator variable is determined. Because of this perfect multicollinearity between the three indicator variables, one of them needs to be dropped from the model. The one that is dropped is called the reference group. We dropped the 0-1 indicator variable signifying membership in the public order crime category. The two remaining indicator variables were coded as follows: violent crime, coded 1 if the inmate was a member of the violent crime group and 0 otherwise; and property crime, coded 1 if the inmate was a member of the property crime group and 0 otherwise. The regression parameter associated with the violent crime category measures the effect of being a member of the violent crime category relative to being a member of the public order crime category, the reference category. Likewise, the regression parameter associated with the property crime category measures the effect of being a member of the property crime category relative to being a member of the public order crime group.

Race had three categories: Black, Other, and White. White was chosen as the reference category. A 0-1 indicator variable for Black and a 0-1 indicator variable for Other were included in the model. Thus, the regression parameter associated with the Black indicator reflects the effect of being Black relative to being White on the disciplinary infraction rate. Likewise, the regression parameter associated with the Other racial group indicator reflects the effect on the disciplinary infraction rate of being a member of the Other racial group compared to the White racial group.

There was a three-category variable representing a combination of prior prison experience and disciplinary infractions during the prior prison experience.

The categories were prior incarceration and at least one infraction, prior incarceration and no infractions, and no prior incarceration. The third category, no prior incarceration, was the reference category. Each of the first two prison experience variables was defined by the appropriate indicator variable.

There was a 0-1 indicator variable representing whether the inmate was on probation or not at the time of his arrest for the current incarceration. Inmates on probation were coded 1, and inmates not on probation were coded 0. Consequently, this variable represents the effect on the infraction rate of being on probation relative to not being on probation. An indicator variable was created for alcohol dependence, with 1 signifying a high risk for alcohol dependence and 0 signifying non-high risk. A similar indicator variable was created for drug dependence.

There were four continuous independent variables: age on admission, in years; time served in North Carolina prisons prior to the study period, in years; years of jail credit applicable to current sentence; and the expected length of the sentence, in years. For these continuous variables, the associated regression parameter reflects the change in the disciplinary infraction rate due to a one-unit increase in the scale (i.e., years) in the associated continuous independent variable.

For this study, the time that each inmate was under observation was not constant. It depended on the time the inmate entered the study and left the study because of release or was no longer observed because the study ended. This varying time period in which the disciplinary infractions were recorded for the inmates needs to be explicitly incorporated in the Poisson regression model. This can easily be done by including an offset in the model. The rate of infractions for a particular inmate can be modeled as $\log_e \frac{\lambda}{t} = \beta_0 + \beta_1 X_1 + \cdots + \beta_p X_p$, where $\frac{\lambda}{t}$ is the rate of infractions for a particular inmate. Time in the denominator is the length of time for which the number of infractions was observed. Because $\log_e \frac{\lambda}{t} = \log_e \lambda - \log_e t$, we can express the previous regression equation as $\log_e \lambda = \log_e t + \beta_0 + \beta_1 X_1 + \cdots + \beta_p X_p$. $\log_e t$ is the offset. It takes a specific value for each inmate, and there is no parameter associated with it to estimate. If we exponentiate both sides of this equation, we have $\lambda = t e^{\beta_0 + \beta_1 X_1 + \cdots + \beta_p X_p}$, and we can write the log likelihood for the Poisson distribution in terms of the specific value of t, the specific values of the X_j's, and the unknown regression parameters. The interpretation of the regression parameters for this model is the same as that for the more conventional model having the same constant observation time for each subject.

We used SAS software (SAS Institute, 2002) to estimate the model. There were 11,738 observations and 14 independent variables, as described previously. The log likelihood for the model was −2661.0231. This value can be compared to the log likelihood for other models in which a set of independent variables was dropped to determine whether the simpler model could predict infraction rates as well as the more complex model (i.e., a likelihood ratio test). The deviance associated with the model was 25659.87. Dividing the deviance by its degrees of freedom yields a measure of the goodness of fit of the model, called the scaled deviance. Because there are 11,723 degrees of freedom (sample size (11,738) − number of estimated regression parameters (15)), the goodness of fit was 2.1888. For the Poisson distribution, the mean is equal to its variance. Under this condition, we would expect the scaled deviance, our measure of goodness of fit, to be approximately 1. If it is smaller than 1, we have a condition called underdispersion; if it is larger than 1, we have overdispersion. In either case, the equal mean and variance requirements of the Poisson distribution are violated, and the goodness of fit of the model is compromised.

This violation of the Poisson distribution assumptions does not have an effect on the regression parameter estimates; it does, however, have an effect on the standard error estimates for the regression coefficients. Overdispersion is usually more common than underdispersion, as in our example. Overdispersion means that the count data associated with the outcome or dependent variable, the infraction counts in our case, are more variable than expected for a Poisson distribution. Thus, the maximum likelihood-based standard error estimates are underestimates of the actual standard errors because the variance of the infraction counts was underestimated by using the Poisson mean equal to variance condition in the likelihood estimating equations. We can correct the original standard errors by multiplying them by the square root of the scaled deviance.

Like most regression models, the most important output is a table with estimated regression parameters, standard errors, t ratios, and significance levels as column headings. Besides the intercept parameter estimate, our Poisson infraction model contained 14 estimated regression parameters. The most important regression parameter was associated with the Structured Sentence (1) versus Fair Sentence (0) indicator variable because this was the main thrust of the study. The remaining 13 independent variables were prisoner background variables that served as control variables to adjust the effect of Structured versus Fair Sentencing for differences in prisoner background characteristics between the two groups. The effects of prisoner background variables on infraction rates were also of interest because they indicated risk factors associated with infraction rates.

We do not present all 14 estimated regression parameters but only those for Structured versus Fair Sentencing, prisoner age, and prior jail and infraction history. Structured versus Fair Sentencing is represented by a single

TABLE 8.1
Parameter Estimates for Poisson Infraction Model

Variable	Parameter Estimate	Standard Error	t Ratio	Significance Level
Structured versus Fair Sentencing	.2413	.0326	7.40	<.0001
Prior jail and infractions versus no prior jail	.5501	.0403	13.65	<.0001
Prior jail and no infractions versus no prior jail	.0413	.0341	1.21	<.2259
Prisoner age	−.0831	.0022	−37.77	<.0001

indicator variable, with 1 indicating Structured Sentencing and 0 indicating Fair Sentencing. Prisoner age is a continuous variable measured in years. Prior jail and infraction history is, as discussed previously, a three-level categorical variable that is summarized by two indicator variables. The first assigns a 1 to prisoners who were previously in jail and committed infractions and a 0 otherwise. The second assigns a 1 to prisoners who were previously in jail but had no infractions and a 0 otherwise. Thus, the first indicator variable represents a contrast between prisoners who were previously in jail and committed infractions and prisoners who were not previously in jail. The second indicator variable represents a contrast between prisoners who were previously in jail but committed no infractions and prisoners with no prior jail experience. Thus, no previous jail time was the reference group for the two indicator variables. The results of the regression analysis are presented in Table 8.1.

Three of the four variables were highly significant predictors of the prisoners' infraction rates. Structure Sentenced prisoners had a significantly higher $\log_e$ (rate of infractions) than Fair Sentenced prisoners. Equivalently, the infraction rate for Structure Sentenced prisoners was higher than that for Fair Sentenced prisoners. Prisoners with prior jail experience and infractions had a higher infraction rate than those with no prior jail experience. The highly significant negative estimated regression parameter, −.0831, associated with prisoner's age indicates that as prisoners get older, their infraction rates decrease.

The estimated regression parameters reflect the additive effect of the associated independent variable on the natural logarithm of the infraction rate. Although the sign and relative magnitude of these estimated regression parameters give one an idea of the effects of the associated variables, the individual estimated regression parameters are difficult to interpret. Consequently, it is common practice to exponentiate the estimated regression parameters (i.e., e^{β}) so that they reflect a multiplicative effect on the infraction rate. Most computer programs will output the exponentiated regression parameter estimates along with a confidence interval whose confidence level can be specified (e.g., 95%).

TABLE 8.2

Infraction Poisson Regression

Variable	Exponentiated Parameter Estimate	95% Confidence Interval
Structured versus Fair Sentencing	1.27	[1.19, 1.36]
Prior jail and infractions versus no prior jail	1.73	[1.60, 1.88]
Prior jail and no infractions versus no prior jail	1.04	[0.97, 1.11]
Prisoner age	0.92	[0.916, 0.924]

Using the estimated regression parameters and standard errors in Table 8.1, we present the exponentiated estimated regression parameters and their 95% confidence intervals in Table 8.2. The original 95% confidence interval is $\hat{\beta} \pm 1.96$ times the standard error of $\hat{\beta}$. Consequently, the confidence interval for $e^{\hat{\beta}}$ is

$$[e^{\hat{\beta}-1.96se_{\hat{\beta}}}, e^{\hat{\beta}+1.96se_{\hat{\beta}}}],$$

where $se_{\hat{\beta}}$ is the standard error of $\hat{\beta}$.

If the additive effect $\hat{\beta}$ is zero, then the multiplicative effect $e^{\hat{\beta}} = e^{0} = 1$. Therefore, a multiplicative effect of 1 signifies no effect. If the multiplicative effect is less than 1, then the associated independent variable has a negative effect on the infraction rate. If the multiplicative is greater than 1, then the associated independent variable has a positive effect on the infraction rate. A negative effect means that as the value of the independent variable increases, the infraction rate decreases, whereas a positive effect means that as the value of the independent variable increases, the infraction rate increases. If the confidence interval covers 1, then the exponentiated regression parameter or, equivalently, the regression parameter is not statistically significant at the given significance level. For example, if a 95% confidence interval covers 1, then the regression parameter is not statistically significant at the .05 level.

The multiplicative effect of Structured versus Fair Sentencing on the total infraction rate is 1.27. This means that the infraction rate for prisoners under Structured Sentencing was 27% higher than that for prisoners under Fair Sentencing. That is, multiply the Fair Sentencing infraction rate by 1.27 to get the infraction rate for Structured Sentencing.

The multiplicative effect of having previously been in jail with an infraction history versus no previous jail is 1.73. The infraction rate for the previous group is 73% higher than that for the latter group. The contrast between the prior jail/no infraction group was not significantly different from the no prior jail group at the .05 level. Notice that the 95% confidence interval contains 1, no multiplicative effect, or, equivalently, no additive effect on the $\log_e$ scale.

The multiplicative effect for age was 0.92, indicating that an increase of 1 year of age has the effect of decreasing the infraction rate by 8%. That is, the effect of age on the total infraction rate is negative. Notice that the additive effect parameter was negative (i.e., $-.0831$). Because age is continuous, we can examine the effect of any arbitrary increase or decrease of age on the infraction rate. For example, an increase of 10 years in age has a multiplicative of $e^{10\hat{\beta}_{age}}$ on the infraction rate because the effect of a 1-year increase is $e^{\hat{\beta}_{age}}$. Because $e^{10\hat{\beta}_{age}}$ equals $(e^{\hat{\beta}_{age}})^{10}$, we simply have to raise $e^{\hat{\beta}_{age}}$, the multiplicative effect on the infraction rate due to a 1-year increase in age, to the 10th power. We can easily do this on any calculator. Because $e^{\hat{\beta}_{age}}$ is 0.92, the multiplicative effect for a 10-year increase in age is $(0.92)^{10}$ or 0.43, a decrease of 57% in the infraction rate.

9. SURVIVAL ANALYSIS

Survival analysis, as the name implies, involves developing regression models for predicting how long an individual or object survives until an event, such as death in the case of an individual or failure in the case of an object such as a machine part, occurs. Survival analysis has wide applicability across all the sciences (Hosmer & Lemeshow, 1999). In medical research, for example, it is used to investigate the impact of various drugs on the survival time of cancer patients. In the physical sciences, it is used to model the time to failure of various subsystems, such as aircraft components. It has also found wide application in the social sciences. For example, it has been used to model the time it takes to find a new job after being laid off, the time a patient spends in a drug abuse treatment program before dropping out, and the time it takes for a prisoner to commit a prison infraction from the time he or she is incarcerated.

For problems such as these, as well as similar problems, we would like to develop regression models to predict survival time or some function of survival time. That is, we would like to determine if a hypothesized set of independent variables or covariates can explain the survival time or the time it takes for an event to occur. For example, it would be useful to know what patient and drug treatment program characteristics are related to length of time a patient spends in the program before voluntarily dropping out. For heroin addicts in methadone programs, it has been found in some studies that patients receiving higher doses of methadone tend to stay in treatment longer. Of course, other variables besides dosage level are included in the models to adjust for other differences, such as gender and age between patients, that may also affect survival time in treatment.

There is an aspect of survival analysis that is not found in the other generalized regression models previously discussed and makes survival analysis more complicated under certain conditions. This complicating aspect is censoring. In many studies of limited duration, survival times may not be available for all individuals. In our methadone treatment example, there may be a relatively large proportion of the patients who may not have dropped out of treatment during the period of the study. We would not have treatment survival times on those patients because they were still in treatment at the end of the study. Another example is a 5-year study of cancer patients in which survival time until death is the outcome variable; some patients may still be living at the end of the study. Although we cannot measure survival time for these patients, we do have some information on these individuals that can be used in estimating the regression parameters in the model. We know that they survived for a certain amount of time. We will see later how this information is used in survival analysis. Besides people who survived to the end of the study, there may be other people who were lost to follow-up so that we do not know their survival time. We may know that they survived up to a certain point in time, however (Figure 9.1).

Survival Time Distributions

Survival time in any study, of course, varies across individuals. It is a continuous random variable and has a density function like any random variable. A number of different distributions are used to characterize survival time, including the Weibull, exponential, gamma, and log-normal. The selection of a particular distribution to use in a survival model depends on the nature of the study, the theoretical justification, and how well the selected distribution fits the empirical study data.

This wide variety of distributions allows a wide variety of shapes and scales (dispersions) of the survival data. Probably the most widely used distribution is the Weibull distribution because it is a two-parameter distribution (a shape parameter, α, and a scale parameter, λ) that allows a wide variety of distributions depending on the value of the parameters, α and λ. Depending on the parameters, the Weibull distribution can approximate the shape of the exponential, gamma, and log-normal distribution. Figure 9.2 plots the Weibull distribution at various values of α but with the same value of λ.

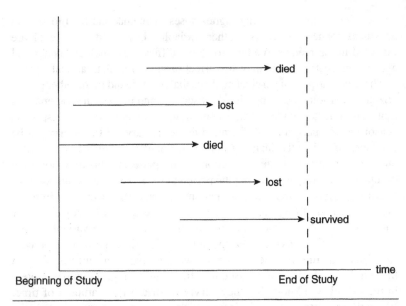

Figure 9.1   Time Lines for Survival Analysis

The simplest survival time distribution is the one-parameter, exponential distribution; it is $f(t) = \lambda e^{-\lambda t}$ with parameter λ. Because it is simple, it is sometimes used in regression models of survival time. The parameters of this regression model, as we shall see, are easy to interpret, but there are some limitations that we will describe later.

Let us now discuss some more concepts related to survival time distributions. Besides characterizing a survival distribution by its density function, $f(t)$, we can also characterize it by its distribution function, $F(t)$. $F(t)$ is the probability that the random variable, T (survival time), is equal to or less than a given value of t. For the reader familiar with calculus, this is the integral

$$F(t) = \int_{0}^{t} f(t)dt.$$

$F(t)$ is the area under the density function $f(t)$ to the left of t (Figure 9.3).

A related function, $S(t)$, is called the survival function. It is the probability that the survival time, T, is equal to or greater than t. Because the area under a density function is 1 and $F(t)$ is the probability that T is less than t, $S(t)$ must be equal to $1 - F(t)$. This can be seen in Figure 9.4. That is, because $F(t)$ is the probability of dying or any other event happening before time t, $S(t)$ must be the probability that the event happens at time t or later (Figure 9.5).

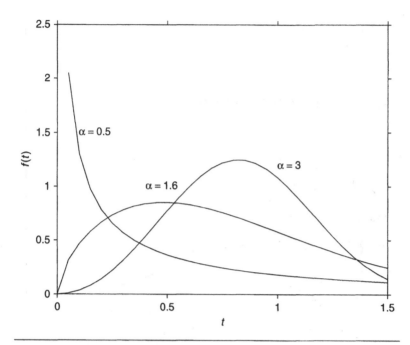

Figure 9.2   Three Weibull Distributions With the Same Scale Parameter ($\lambda = 1.2$) But Different Shape Parameters

A function of time that is of major interest in survival analysis is the hazard function, $h(t)$. The hazard function $h(t)$ is defined as $f(t)/S(t)$ and can be interpreted as the instantaneous probability that the event happens at time t given that the individual has survived to time t. It is a conditional probability and is the ratio of density function $f(t)$ to the survival function $S(t)$. Like the density function, the hazard function can take many forms, as shown in Figure 9.6.

Exponential Survival Model

The simplest parametric survival analysis model involves the exponential distribution. The density function, as discussed previously, is $f(t) = \lambda e^{-\lambda t}$ and involves a single parameter λ. The corresponding survival function is $e^{-\lambda t}$ so that the hazard function

$$h(t) = \frac{f(t)}{S(t)} = \frac{\lambda e^{-\lambda t}}{e^{-\lambda t}} = \lambda.$$

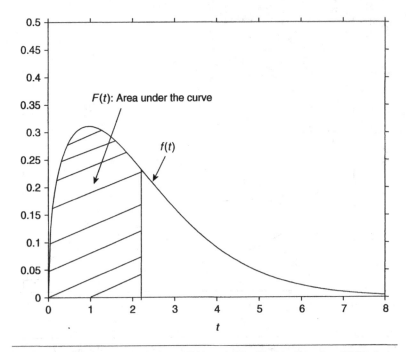

Figure 9.3   Density Function ($f(t)$) and Distribution Function ($F(t)$) for a Survival Time Random Variable

Thus, the hazard is a constant and not a function of time, as it is in more complex two-parameter distributions such as the Weibull. Because the hazard, λ, is always positive, we model $\log_e \lambda$ as a linear function of the independent variables. That is, $\log_e h(t) = \log_e \lambda = \beta_0 + \beta_1 X_1 \cdots +\beta_p X_p$ or, equivalently, $\lambda = e^{\beta_0 + \beta_1 X_1 \cdots + \beta_p X_p}$. For uncensored data, we can substitute these in the log-likelihood function and solve for the maximum likelihood estimates of the regression parameters. That is, the contribution to the log likelihood of the ith observation is $\log_e f(t_i) = \log_e \lambda_i - \lambda_i t_i = \beta_0 + \beta_1 X_{1i} + \cdots + \beta_p X_{pi} - e^{\beta_0 + \beta_1 X_{1i} + \cdots \beta_p X_{pi}} t_i$.

If the data are censored, then the contribution of each observation to the likelihood function depends on whether the observed value of t_i for the ith observation is censored or not. For uncensored survival times, the contribution to the log likelihood is $\log f(t_i)$, as previously shown. For censored survival time, $f(t_i)$ is not appropriate to use in the likelihood function. For censored observations, however, we know that the survival time was at least t_i even if we do not know the exact survival time. This is still useful information about the regression parameters to use in the likelihood function. Thus, for censored

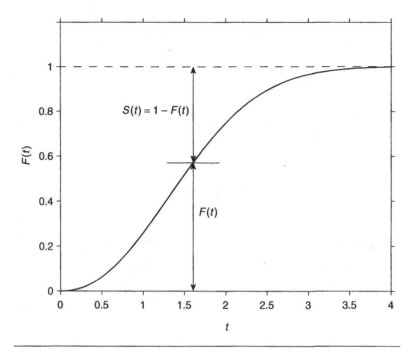

Figure 9.4   Distribution, $F(t)$, and Survival Distribution, $S(t)$

observation, we use $S(t_i)$ as the contribution to the likelihood function. If we use a censoring indicator, δ_i, for each observation such that $\delta_i = 1$ if t_i is uncensored and $\delta_i = 0$ if t_i is censored, then the likelihood function can be written as

$$\prod_i f(t_i)^{\delta_i} S(t_i)^{1-\delta_i}$$

or

$$\prod_i \lambda_i^{\delta_i} e^{-\lambda_i t_i},$$

where $\prod_i$ indicates the product of the individual likelihoods. The log likelihood is $\sum_{i=1}^{n} \delta_i \log \lambda_i - \lambda_i t_i$. Substituting $\beta_0 + \beta_1 X_{1i} + \cdots + \beta_p X_{pi}$ for $\log_e \lambda_i$ and $e^{\beta_0 + \beta_1 X_{1i} + \cdots \beta_p X_{pi}}$ for λ_i in the log likelihood and taking the partial derivatives with respect to each of the regression parameters, we obtain the set of maximum likelihood estimating equations. Because the equations are nonlinear, we use an iterative numerical algorithm to estimate the regression parameters and their standard errors.

58

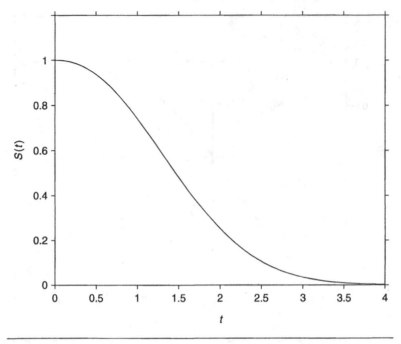

Figure 9.5   Survival Distribution, $S(t)$

Like all generalized linear models, we can compare the fit of various models by comparing the difference in deviances or, equivalently, use the likelihood ratio test, which is $\chi^2 = -2[\ell(\hat{\beta}_1, 0) - \ell(\hat{\beta}_1, \hat{\beta}_2)]$, where $\ell(\hat{\beta}_1, 0)$ is the log likelihood of the simpler model with β_2 set to 0, which is nested within the more complex model, whose log likelihood is $\ell(\hat{\beta}_1, \hat{\beta}_2)$. The more complex model contains the additional set regression parameters, β_2. The degree of freedom for the chi-square is equal to the difference in the number of parameters between the simple and the complex model. If the chi-square is statistically significant, then the simpler model is rejected in favor of the more complex model. If the nested models differ only by one parameter, then we can use the estimated regression parameters and their associated standard errors from the full model and conduct a t test to determine if a particular variable needs to be in the model.

Because $\lambda_i = e^{\beta_0 + \beta X_{1i} + \cdots \beta_p X_{pi}}$, the effect of each independent variable is multiplicative rather than additive, as in the classical multiple regression model. For a positive regression parameter β_j, a one-unit increase in the associated independent variable X_j increases the hazard, λ, by a factor of e^{β_j}. If β_j is negative, then each one-unit increase in X_j decreases the

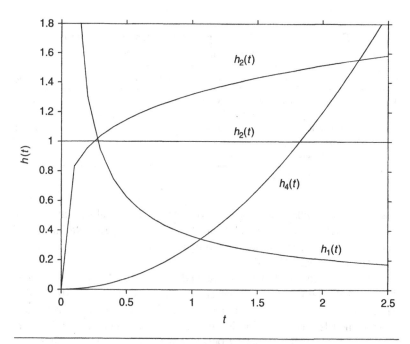

Figure 9.6   Hazard Function for the Weibull Distribution

hazard by a factor of $1 - e^{\beta_j}$. A two-unit increase in X_j for a positive β_j increases the hazard by $e^{2\beta_j}$ and so on. Likewise, a two-unit increase in X_j for a negative β_j decreases the hazard by $1 - e^{2\beta_j}$.

It is interesting to note that the mean of the exponential distribution is $1/\lambda$, the inverse of the hazard. Therefore, as the hazard for the exponential distribution increases, the mean decreases and vice versa. Thus, we can use the estimated regression parameters to model either the hazard or the mean of the exponential distribution for the ith observation. These are two ways of examining the same process.

Example of Exponential Survival Model

A classic example of the application of the exponential survival model is the prediction of time to onset of leukemia for a group of children who were in remission (Breslow, 1974). There were three predictor or prognostic variables: common log of the white blood count (log WBC), age, and age squared. The log WBC was used to compensate for the skewness or outlying values of WBC. Age squared was used because previous research showed that survival is longest for children in the middle age range (i.e., age

TABLE 9.1

Regression Coefficient, Log Likelihood, and Chi-Square
for Exponential Survival Models

Model No.	Prognostic Variables	Log Likelihood	Regression Coefficient(s)	χ^2	df
1	None (intercept only)	−1332.925			
2	log WBC	−1316.999	.72	31.85	1
3	log WBC	−1314.065	.67	37.72	3
	Age		−.14		
	Age squared		.011		

SOURCE: Adapted from Breslow (1974).

has a curvilinear relationship with survival time). The results of fitting three models—intercept only (no covariates), single prognostic variable, and WBC—and the other with all three prognostic factors (WBC, age, and age squared) are presented in Table 9.1.

The chi-square for Model 2 is obtained by −2[log likelihood for Model 1 − log likelihood for Model 2], which is −2[−1332.925 − (−1316.999)] = 31.85; this is also equal to the difference in the deviances between the full and reduced models. The chi-square has one degree of freedom because it differs from the intercept-only model by only one regression parameter. Thus, it is highly significant at the .001 level. Chi-square with one degree of freedom is approximately equal to t^2 when the number of observations is large. Thus, the t value associated with a chi-square of 31.85 is $\sqrt{31.85} = 5.64$, and we know that a t of approximately 2.00 is significant at the .05 level.

The chi-square for Model 3 is associated with age, and age squared added to log WBC is also highly significant ($\chi^2 = 37.72$). We would expect this model to be significant because it contains log WBC, which by itself is a highly significant predictor. The question is whether adding age and age squared provides a significantly better fit than log WBC alone. We can test the null hypothesis that the regression coefficients associated with age and age squared are both zero by taking two times the difference in the log likelihood for the two models. Referring the resulting chi-square of 5.87 to a chi-square table with two degrees of freedom, we find that it is not significant at the .05 level. Thus, we accept the null hypothesis that the parameters associated with age and age squared are zero. There are two degrees of freedom because the difference in the number of parameters between the two models is two. Thus, we accept the simpler WBC model. Again, 5.87 is also equal to the difference in the deviances for the two models.

The positive parameter associated with log WBC was highly significant and positive (.72). This indicates an increasing hazard as WBC increases or, equivalently, a decreasing expected survival time. This exponential survival model is an example of a proportional hazards model. The hazard function that is constant for the exponential distribution varies as a function of the independent variables but retains the same shape—a horizontal line. For any proportional hazards model, the importance of the individual covariates is given by e^{β_j}, where β_j is the regression parameter associated with the jth covariate. The value of e^{β_j} represents the multiplicative change in the hazard associated with a one-unit increase in the associated covariate X_j holding all other covariates constant. In our example, the change due to a one-unit increase in log WBC was $e^{0.72}$ or 2.05, indicating an approximate doubling of the hazard.

CONCLUSIONS

Generalized linear models provide the flexibility to handle a wide range of data with different probability distributions for the dependent variable. They are generalizations of the classical regression model in which the dependent variable is assumed to have a normal distribution and the conditional mean is assumed to be a linear function of the independent variables. Generalized linear models assume that a function of the mean, $g(\mu_i)$, rather than the mean, is a linear function of the independent variables. The dependent variable can have any number of distributions from the family of exponential distributions. The link function that links the mean to the linear predictor is often dictated by the particular error distribution of the dependent variable. This is called the canonical link. For the normal distribution, it is the identity link, $g(\mu_i) = \mu_i$; for the binomial distribution, it is the logit link $g(\mu_i) = \log_e(\frac{\mu_i}{1 - \mu_i})$; and for the Poisson distribution, it is the log link, $g(\mu_i) = \log_e(\mu_i)$. For the proportional hazards model, the log of the hazard is a linear function of the independent variables. Deviance is used to assess the goodness of fit of a particular model. The difference in deviances can be used to compare the fit of two alternative nested models.

The generalized models discussed in this book assumed that the observations are independent. The models can be extended to the case in which the observations are correlated due to the clustering of the observations within a higher-level unit, such as schools, clinics, and classes. These models are called mixed effect (or random effect) models because they contain a fixed component involving the regression parameter, like the models we discussed, plus a random component representing the effect of clusters. The random component accounts for the correlation of the observations within the clusters.

There are many other generalized linear models that we have not discussed, but they are less frequently used than the ones discussed herein.

APPENDIX

The analyses reported in this book were performed by the statistical package called SAS (SAS Institute, 2002). There are many procedures available in SAS that can perform a wide range of statistical analyses. Here, we summarize some of the relevant SAS procedures for fitting generalized linear models.

One very flexible procedure called **PROC GENMOD** in SAS can fit all the generalized linear models discussed in this book. There are nine built-in link functions, including identity, log, logit, and probit; it also allows users to specify their own link functions. Combined with the seven built-in distributions, including binomial, gamma, inverse Gaussian, multinomial, negative binomial, normal, and Poisson, it opens many new opportunities for statistical modeling. We provide some examples of how to use this procedure. In these examples, user-specified argument is printed in italics, DV stands for dependent variable, and IV stands for independent variable in the model. Remarks for the commands are denoted as /* ... */. Notice that each SAS command statement has to be ended with a semicolon (;).

(a) Linear regression: Link function is identity and distribution is normal.
proc genmod data = *name of the data set* ;
　class *name of the categorical independent variable if exists* ;
　model *name of the DV* = *name of the IVs* / **dist** = **normal**
　　　　　　　　　　　　　　　　　　　　　link = **identity;**
run ;

The regression analysis reported in Table 3.1 can be obtained using the following commands:

/* The following commands are for reading data into SAS */
　data table31 ;
　input *y x1 x2 x3 x4 x5 x6* ;
　datalines ;
5 2 3 4 5 5 6 /* This is subject 1's data, the seven columns contain the value
　　　　for the */
　　　　/* seven variables in the order defined in the **'input'** statement. */
. . . .
;
/* We do not need **'class'** statement below because all the IVs are continuous */
/* variables. */

```
proc genmod data = table31 ;
model y = x1 x2 x3 x4 x5 x6 / dist = normal link = identity ;
run;          /* 'run' is to execute the above SAS statements. */
```

(b) Logistic regression: Link function is logit and distribution is binomial.

```
proc genmod data = name of the data set ;
   class name of the categorical independent variable if exists ;
   model name of the DV = name of the IVs / dist = binomial
                                              link = logit ;
run ;
```

The logistic regression analysis reported in Table 7.1 can be obtained using the following commands:

```
/* The following commands are for reading data into SAS */
   data table71 ;
   input interv JLLV JVC YM MI ;
   label interv = 'intervention (1) or not (0)'
         JLLV = 'juvenile liquor law violations'
         JVC = 'juvenile arrest rate for violent crime (no. of arrests
                per 1000 juveniles)'
         YM = 'percentage of young males as a percent of the country
              population'
         MI = 'median income for the country measured in dollars (in
              thousands)'
   ;
   datalines ;
1    250   0.64   0.25   11.5
. . . . . .
   ;
```

```
proc genmod data = table71 ;
   model interv = JLLV JVC YM MI / dist = binomial link = logit ;
run ;
```

(c) Poisson regression: Link function is log and distribution is Poisson.

```
proc genmod data = name of the data set ;
   class name of the categorical independent variable if exists ;
   model name of the DV = name of the IVs / dist = poisson link = log
                     offset = name of the variable used as an offset ;
run ;
```
Notice that the offset variable cannot be the DV or IV.

The Poisson regression analysis reported in Table 8.1 can be obtained using the following commands:

```
data table81 ;
    input infract sentence prior1 prior2 age time ;
    log_time = log(time); /*This command creates the offset variable */
    label infract = 'number of disciplinary infraction'
        sentence = 'structured sentencing (1) vs. fair sentencing (0)'
        prior1 = 'prior jail and infractions (1) vs. no prior jail (0)'
        prior2 = 'prior jail and no infractions (1) vs. no prior jail (0)'
        age = 'age of the prisoner'
        time = 'length of time for which the number of infractions were
                observed'
;
datalines ;
4   1   1   0   24   5
. . . . . .
;

proc genmod data = table81 ;
    class sentence prior1 prior2 ;
    model infract = sentence prior1 prior2 age / dist = binomial
                                            link = logit
                                            offset = log_time ;
run ;
```

(d) Survival regression: **PROC GENMOD** does not analyze censored data or provide other useful lifetime distributions such as the Weibull or log-normal. It can be used, however, for modeling uncensored data with the gamma distribution, and it can provide a statistical test for the exponential distribution against other gamma distribution alternatives.

```
proc genmod data = name of the data set ;
    class name of the categorical independent variable if exists ;
    model name of the DV = name of the IVs / dist = gamma link = log ;
run ;
```

Exponential survival regression can be fitted by adding a subcommand **SCALE**:

```
proc genmod data = name of the data set ;
    class name of the categorical independent variable if exists ;
    model name of the DV = name of the IVs / dist = gamma link = log
                                            scale = 1 ;
run ;
```

The exponential survival analysis reported in Table 9.1 can be obtained using the following commands:

```
data table91 ;
input onset age wbc ;
age_sq = age*age;    /*This command creates the squared term for age */
log_wbc = log10(WBC) ;    /*This command creates the log₁₀(WBC) */
label onset = 'onset time of leukemia'
        wbc = 'white blood count (thousand per microliter)'
      age = 'age' ;
datalines ;
8    12    14
... ...
;


/* Model 1 */
proc genmod data = table91 ;
   model onset = / dist = gamma
                   link = log
                   scale = 1 ;
run ;


/* Model 2 */
proc genmod data = table91 ;
   model onset = log_wbc / dist = gamma
                          link = log
                          scale = 1 ;
run ;


/* Model 3 */
proc genmod data = table91 ;
   model onset = log_wbc age age_sq / dist = gamma
                                     link = log
                                     scale = 1 ;
run ;
```

To obtain the predicted values, Pearson and deviance residuals described at the end of Chapter 6, we can use the subcommand **OUTPUT** to output these values for assessing the goodness of fit of the model. For illustration, consider the Poisson regression example,

```
proc genmod data = name of the dataset ;
   class name of the categorical independent variable if exists ;
   model name of the DV = name of the IVs / dist = poisson link = log
      offset = name of the variable used as an offset ;
      output out = user-specified output file name
         pred = user-specified variable name for predicted values
      reschi = user-specified variable name for Pearson residuals
      resdev = user-specified variable name for deviance residuals ;
```

To print out the predicted values and the residuals,

proc print data = *output file name from the output subcommand above* ;

PROC GENMOD can also be used to analyze longitudinal or clustered data, which we do not discuss in this book because of space constraints.

There are also other SAS procedures available to fit special classes of generalized linear models. We list some commonly used ones for readers' reference.

For Linear Regression

PROC REG and PROC GLM*: Both procedures can fit linear regression and allow categorical and continuous independent variables. For **REG** procedure, users have to create the dummy variables to represent the categorical independent variable. For **GLM** procedure, users do not have to do so but only need to declare the categorical independent variables in the 'CLASS' statement, a subcommand of the GLM procedure.
*GLM here stands for general linear models.

For Logistic Regression

PROC LOGISTIC: It can fit logistic regression models for binomial and ordinal outcomes. It also provides a wide variety of model-building methods and computes numerous regression diagnostics.

For Probit Regression

PROC PROBIT: It performs logistic regression, ordinal logistic regression, as well as probit regression. The **PROBIT** procedure is useful when the dependent variable is either dichotomous or polychotomous, and the independent variables are continuous. Probit regression is similar to logistic regression, except the link function is normal (Gaussian) instead of logit.

For Survival Regression

PROC PHREG: It performs regression analysis of survival data based on the Cox's proportional hazards model, which assumes a parametric form for the effects of the explanatory variables but an unspecified form for the underlying survivor function. It also allows for censored survival time observations, which cannot be handled in **PROC GENMOD**. If there are interval-censored observations (exact survival times are not observed but only known up to an interval), **PROC LIFTEST** can be used instead.

There is a newly developed procedure called **PROC QLIM** that can fit both univariate and multivariate logit and probit models.

REFERENCES

Breslow, N. (1974). Covariance analysis of survival data under the proportional hazards model. *International Statistics Review, 43*, 43-54.

Chatterjee, S., & Price, B. (1977). *Regression analysis by example.* New York: John Wiley.

Fahrmeir, L., & Tutz, G. (1994). *Multivariate statistical modeling based on generalized linear models.* Berlin: Springer-Verlag.

Hosmer, D. W., & Lemeshow, S. (1999). *Applied survival analysis.* New York: John Wiley.

Le, C. T. (1998). *Applied categorical data analysis.* New York: John Wiley.

Lee, E. T. (1992). *Statistical methods for survival data analysis* (2nd ed.). New York: John Wiley.

McCullagh, P., & Nelder, J. A. (1989). *Generalized linear models* (2nd ed.). London: Chapman & Hall.

McCulloch, C. E., & Searle, S. R. (2001). *Generalized, linear, and mixed models.* New York: John Wiley.

Nelder, J. A., & Wedderburn, T. W. M. (1972). Generalized linear models. *Journal of the Royal Statistical Society (Series A), 135*, 370-384.

SAS Institute, Inc. (2002). *SAS/STAT 9 user's guide* (Vols. 1-3). Cary, NC: Author.

INDEX

ABOUT THE AUTHORS

George H. Dunteman, Ph. D., was Chief Scientist at the Research Triangle Institute, where he was actively involved in applied research, primarily in the social and behavioral sciences, when he passed away in April, 2004. He had previously held research appointments at the Educational Testing Service and the U.S. Army Research Institute. He had also held assistant and associate professorships at the University of Rochester and the University of Florida, respectively. During the 1987-1988 academic year, he was a visiting professor of management in the Babcock School of Management at Wake Forest University, where he taught the MBA core course in quantitative methods. He was on the editorial board of *Educational and Psychological Measurement* and had published widely in professional journals. He previously authored three books for Sage: *Introduction to Linear Models* (1984); its companion volume, *Introduction to Multivariate Analysis* (1984); and No. 69 in this series, *Principle Components Analysis* (1989).

Moon-Ho R. Ho, Ph. D., is Assistant Professor of Psychology at McGill University in Canada and Assistant Professor in the School of Humanities and Social Sciences at Nanyang Technological University and the Center for Research in Pedagogy and Practice at the National Institute of Education in Singapore, where he teaches advanced undergraduate and graduate-level statistics courses. His major areas of quantitative interest include analysis of brain imaging data, multilevel modeling, nonstationary time series data analysis method, resampling techniques, social network analysis, and structural equation modeling. His substantive interests include cognitive models for judgment and decision making, attitude formation, mechanisms underlying changes in health-related behavior, and father involvement in child rearing.